III Punctuation

IV Spelling and mechanics

V Research and documentation

VI Special writing situations

Finding What You Need

This reference contains essential information for writing: the writing process, usage, grammar, punctuation, research writing, source citation, and special writing situations such as oral presentations and business writing.

The handbook provides many ways to reach its information:

Use a table of contents.

Inside the front cover, a brief contents gives an overview of the handbook. Inside the back cover, a detailed outline lists all the book's topics.

Use the index.

At the end of the book, this alphabetical list includes all topics, terms, and problem words and expressions.

Use the glossaries.

The Glossary of Usage (p. 219) clarifies words and expressions that are often misused or confused, such as *hopefully* or *affect/effect*. The Glossary of Terms (p. 229) defines grammatical terms, including all terms marked ° in the text.

Use the elements of the page.

❶ Running head (header) showing the topic being discussed on this page.
❷ Chapter number and title.
❸ Tip for computer use.
❹ Section heading containing a main topic or convention. Heading code consists of chapter number (**19**) and heading letter (**a**).
❺ Section subheading.
❻ Small raised circles indicating terms defined in the Glossary of Terms (p. 229).
❼ Page tab containing the code of the last heading on the page (**19a**).
❽ Examples, always indented, often showing revision.

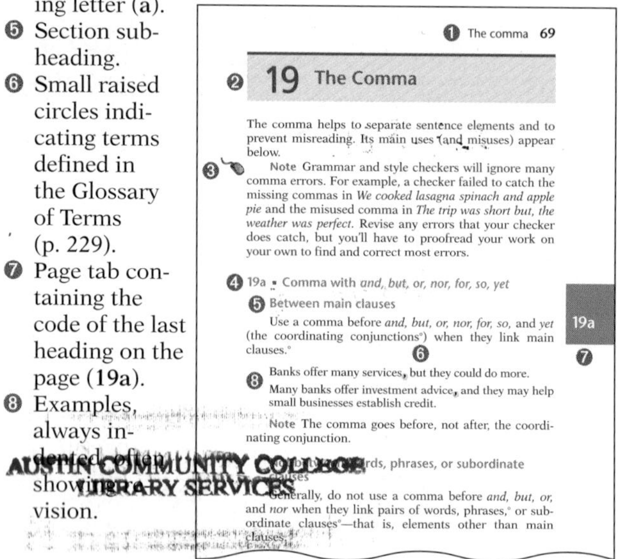

❶ The comma 69

❷ 19 The Comma

The comma helps to separate sentence elements and to prevent misreading. Its main uses (and misuses) appear below.

❸ Note Grammar and style checkers will ignore many comma errors. For example, a checker failed to catch the missing commas in *We cooked lasagna spinach and apple pie* and the misused comma in *The trip was short but, the weather was perfect.* Revise any errors that your checker does catch, but you'll have to proofread your work on your own to find and correct most errors.

❹ 19a Comma with *and, but, or, nor, for, so, yet*

❺ Between main clauses

Use a comma before *and, but, or, nor, for, so,* and *yet* (the coordinating conjunctions°) when they link main clauses.° ❻

19a ❼

❽ Banks offer many services, but they could do more.
Many banks offer investment advice, and they may help small businesses establish credit.

Note The comma goes before, not after, the coordinating conjunction.

Not between words, phrases, or subordinate clauses

Generally, do not use a comma before *and, but, or,* and *nor* when they link pairs of words, phrases,° or subordinate clauses°—that is, elements other than main clauses.

FIFTH EDITION

The Little, Brown Essential Handbook

Jane E. Aaron

Acquisitions Editor: Brandon Hight
Development Editor: Carol Hollar-Zwick
Executive Marketing Manager: Megan Galvin-Fak
Senior Supplements Editor: Donna Campion
Managing Editor: Bob Ginsberg
Project Coordination, Text Design, and Electronic Page Makeup: Nesbitt Graphics, Inc.
Senior Cover Design Manager: Nancy Danahy
Cover Design Manager: Wendy Ann Fredericks
Cover Designer: Kay Petronio
Manufacturing Manager: Mary Fischer
Senior Manufacturing Buyer: Alfred C. Dorsey
Printer and Binder: Quebecor, Taunton
Cover Printer: Phoenix Color Corporation

Anderson, Daniel. Screen shot, "The Changing Face of College Computing," *http://sites.unc.edu/~daniel/changing_face/index.html.* Copyright © Daniel Anderson. Used with permission.
EBSCO Publishing. Screen shot of *EBSCOhost Academic Search* used with permission from EBSCO Publishing.
Google. Screen shots used with permission from Google. GOOGLE is a trademark of Google Inc.

Library of Congress Cataloging-in-Publication Data

Aaron, Jane E.
The Little, Brown essential handbook / Jane E. Aaron.— 5th ed.
p. cm.
Rev. ed. of: The Little, Brown essential handbook for writers. 4th ed. c2003.
Includes index.
ISBN 0-321-33159-1
1. English language—Grammar—Handbooks, manuals, etc. 2. English language—Rhetoric—Handbooks, manuals, etc. 3. Report writing—Handbooks, manuals, etc. I. Aaron, Jane E. Little, Brown essential handbook for writers. II. Title.
PE1112.A24 2005
808'.042—dc22

2005001807

se visit our website at *http://www.ablongman.com/littlebrown.*

0-321-33159-1

78910—QWT—08070605

Introduction: The Process of Writing

I1

Like most writers (even very experienced ones), you may find writing sometimes easy but more often challenging, sometimes smooth but more often halting. Writing involves creation, and creation requires freedom, experimentation, and, yes, missteps and midcourse corrections. Instead of proceeding in a straight line over a clear path, you will likely experience the recursive nature of writing. You might start writing without knowing what you have to say, circle back to explore a new idea, or keep going even though you're sure you'll have to rewrite later.

As uncertain as the writing process may be, you can bring some control to it by assessing your writing situation and by anticipating the stages you may cycle through.

I1 ▪ The writing situation

The **writing situation** consists of the requirements and the options that determine what and how you will write. Considering your situation at the start of a project will tell you a great deal about how to proceed.

Subject, audience, and purpose

Whenever you write to be read by others, you are communicating something about a topic to a particular audience of readers for a specific reason.

Subject

- What does your writing assignment tell you to do? If you don't have an assignment, what do you want to write about?
- What interests you about the subject? What do you already have ideas about or want to know more about? What is your attitude toward the subject: serious, angry, puzzled, amused?
- Is your subject limited enough so that you can cover it well in the space and time you have?

Audience

- Who will read your writing? What can you assume your readers already know and think about your

Visit *www.ablongman.com/littlebrown* for added help and exercises on the process of writing.

subject? How can your assumptions guide your writing so that you tell readers neither too little nor too much?

- Do your readers have any characteristics—such as educational background, experience in your field, or political views—that could influence their reception of your writing?
- What is your relationship to your readers? How formal or informal should your writing be?
- What do you want readers to do or think after they read your writing?

Purpose

- What aim does your assignment specify? For instance, does it ask you to explain something or argue a point?
- Why are you writing? What do you want your work to accomplish?
- How can you best achieve your purpose?

Research

- What kinds of evidence—such as facts, examples, and the opinions of experts—will best suit your subject, audience, and purpose?
- Does your assignment require you to consult sources of information or conduct other research, such as interviews, surveys, or experiments?
- Besides the requirements of the assignment, what additional information do you need to develop your subject? How will you obtain it?
- What style should you use to cite your sources? (See pp. 141–42 on source documentation in the academic disciplines.)

An overview and checklist of the research-writing process appears on page 104.

ESL Research serves different purposes in some other cultures than it does in the United States. For instance, students in some cultures may be expected to consult only well-known sources and to adhere closely to the sources' ideas. In US colleges and universities, students are expected to look for relevant and reliable sources, whether well known or not, and to use sources mainly to support their own ideas.

Deadline and length

- When is the assignment due? How will you apportion the work you have to do in the available time?
- How long should your writing be? If no length is assigned, what seems appropriate for your topic, audience, and purpose?

Document design

- What format does the assignment require? (Several academic formats are discussed in this book: MLA, p. 159; APA, p. 175; and Chicago, p. 189. See also p. 211 on business-writing formats.)
- Even if a particular format is not required, how might you use margins, headings, illustrations, and other elements to achieve your purpose? (See pp. 200–05.)

I2 ▪ Development and focus

With a sense of your writing situation, you can begin to develop a draft of your paper. The guidelines below appear as stages to emphasize the kinds of thinking involved at various points in composing. But the stages are not fixed: you will inevitably circle back through them as your paper evolves, and you may find that some other sequence helps you touch the same bases.

Exploration

Most writers start a project by exploring ideas about their subject.

- In an informal document—perhaps a brainstorming list or a branching diagram—try to discover and gather your own thoughts on your subject. Concentrate on opening up avenues: delay editing your ideas until you've seen how they come out in writing.
- Begin finding and evaluating appropriate sources of information and opinion to support and extend your own ideas. (See pp. 111–27 for tips on research.)

Thesis

The **thesis** is the central idea of a piece of writing: the entire work develops and supports that idea. Though sometimes unstated, a thesis should always govern a paper; usually it appears in a **thesis statement** somewhere in the paper. To develop a thesis statement, try the following:

- Focus your thoughts and information on a single dominant question you seek to answer. Let the question guide your planning and research, but also allow it to change as information and ideas accumulate.
- As your work proceeds, begin answering your question in a one- or two-sentence statement of your main idea. Like your question, this thesis statement may change as your ideas do, but eventually it will be the focus of your final paper.

Here are pairs of questions and answering thesis statements from various disciplines:

Literature question	What makes the ending of Kate Chopin's "The Story of an Hour" believable?
Thesis statement	The ironic ending of "The Story of an Hour" is believable because it is consistent with the story's other ironies.
History question	How did eviction from its homeland in 1838 affect the Cherokee Nation of Native Americans?
Thesis statement	Disastrous as it was, the forced resettlement of the Cherokee did less to damage the tribe than did its allegiance to the Confederacy during the Civil War.
Psychology question	How common is violence between partners in dating relationships?
Thesis statement	The survey showed that violence may have occurred in a fifth of dating relationships among students at this college.
Biology question	Does the same physical exertion have the same or different effects on the blood pressure of men and women?
Thesis statement	After the same physical exertion, the average blood pressure of female participants increased significantly more than the average blood pressure of male participants.

Notice that these thesis statements are concise, are limited to a single idea, and offer a specific opinion that will be supported in the paper.

ESL In some other cultures it is considered unnecessary or impolite for a writer to express an opinion or to state his or her main idea outright. When writing in English, you can assume that your readers expect a clear idea of what you think.

Plan

Even with a short piece of writing such as a letter or memo, your work will proceed more smoothly when you have a plan for it. You may need nothing more than a scratch outline that lists your major points, perhaps with the most significant support for each point.

Most essays or papers divide into three parts:

- The **introduction**—usually a paragraph or two—presents the topic, sometimes provides background, narrows the topic, and often includes the thesis statement.

- The **body**, the longest part, contains the substance of the paper, developing parts of the thesis. See "Paragraphs" below.
- The **conclusion**—usually a paragraph—ties together the parts of the body, sometimes restating the thesis, summarizing the major points, suggesting implications of the thesis, or calling for action.

ESL If English is not your native language, you may not be accustomed to the pattern of introduction-body-conclusion. For instance, instead of focusing the introduction quickly on the topic and thesis, writers in your native culture may take an indirect approach. And instead of arranging body paragraphs to give general points and then the evidence for those points, writers in your native culture may leave the general points unsupported (assuming that readers will supply the evidence themselves) or may give only the specifics (assuming that readers will infer the general points). When writing in English, you need to address readers' expectations for directness and for the statement and support of general points.

Paragraphs

The body of your paper will consist of paragraphs that develop the major points contributing to the thesis. A point may require a single paragraph or two or three paragraphs. Generally, body paragraphs have their own structures:

- **A topic sentence** (often the first or second sentence) states the point that the paragraph develops.
- The other sentences offer examples, facts, expert opinions, and other evidence to support the topic sentence.
- Occasionally, a concluding sentence ties the evidence together or prepares for the point of the next paragraph.
- To bind paragraphs and sentences so that they flow smoothly, you can use **transitional expressions** such as *first, however,* and *in addition.* (See p. 239 for a list.)

I3 ▪ Revision

When you have a draft of your paper, let it rest for a while to get some distance from it and perhaps to gather comments from others, such as your classmates or instructor. Then revise the draft against the checklist on the next page, concentrating on the effectiveness of the whole. (Leave style, correctness, and other specific issues for the separate step of editing, discussed on p. 7.)

REVISION CHECKLIST

- ☑ **Purpose** What is the paper's purpose? Does it conform to the assignment? Will it be clear to readers?
- ☑ **Thesis** What is the thesis? Does the paper demonstrate it? What does each paragraph and each sentence contribute to the thesis? If there are digressions, should they be cut or reworked?
- ☑ **Organization** What are the major points supporting the thesis? (List them.) How effective is their arrangement for the paper's purpose? Does the paper flow smoothly so that readers will follow easily?
- ☑ **Development** How well do facts, examples, and other evidence support each major point and the thesis as a whole? Will readers find the paper convincing?
- ☑ **Tone** Is the paper appropriately formal or informal for its readers? Does it convey your attitude appropriately—for instance, is it neither too angry nor too flippant?
- ☑ **Use of sources** Have you used sources to support, not substitute for, your own ideas? Have you integrated borrowed material into your own sentences? (See pp. 127–36.)
- ☑ **Title, introduction, and conclusion** Does the title convey the paper's content accurately and interestingly? Does the introduction engage and focus readers' attention? Does the conclusion provide a sense of completion?
- ☑ **Format** Does the format of your paper suit your purpose and your audience's likely expectations for such papers?

Note If you write on a computer, print your draft and revise it on paper. You'll be able to view whole sections at once, and the different medium can help you see flaws you may have missed on screen.

I4 ▪ Source documentation

If you draw on outside sources in writing your paper, you must clearly acknowledge those sources:

- See pages 136–41 for advice on when to acknowledge sources so that you avoid even the appearance of plagiarism.
- See pages 141–42 on documenting sources. This book covers four documentation styles: MLA for English and many other humanities (pp. 142–59), APA for psychology and some other social sciences (pp. 163–75), Chicago for history and other humanities (pp. 178–89), and CSE for many natural sciences and mathematics (pp. 191–98).

ESL Cultures have varying definitions of a writer's responsibilities to sources. In some cultures, for instance, a writer need not cite sources that are well known. In the United States, in contrast, a writer is obligated to cite all sources. (See also p. 2 on research.)

I5 ▪ Editing and proofreading

Much of this book concerns editing—tightening or clarifying sentences, polishing words, repairing mistakes in grammar and punctuation. Leave this work until after revision so that your content and organization are set before you tinker with your expression. For editing guidelines, see the checklists on pages 10 (effective sentences), 32 (grammatical sentences), 68 (punctuation), and 90 (spelling and mechanics).

Most writers find that they spot errors better on paper than on a computer screen, so edit a printout if you can. And be sure to proofread your final draft before you submit it, even if you have used a spelling checker or similar aid (see below).

Spelling checkers

A spelling checker can be a great ally: it will flag words that are spelled incorrectly and usually suggest alternative spellings that resemble what you've typed. However, this ally has limitations:

- *A spelling checker will not flag misused words.* It cannot recognize typos such as *of* for *or, form* for *from*, or *now* for *not*. Nor can it spot misuses of commonly confused words such as *their/there, its/it's,* and *affect/effect.*
- *The checker may flag a word that you've spelled correctly*, just because the word does not appear in its dictionary.
- *The checker may suggest incorrect alternatives.* Before accepting any highlighted suggestion from the checker, you need to verify that the word is actually what you intend. Consult an online or printed dictionary when you aren't sure of the checker's recommendations.

In the end *the only way to rid your papers of spelling errors is to proofread your papers yourself.*

For more advice on spelling, see Chapter 26.

Grammar and style checkers

Grammar and style checkers can flag incorrect grammar or punctuation and wordy or awkward sentences. You may be able to customize a checker to suit your needs and habits as a writer—for instance, instructing it to look for problems with subject-verb agreement or for passive verbs.

Like spelling checkers, however, grammar/style checkers are limited:

- *They miss many errors* because they are not yet capable of analyzing language in all its complexity.
- *They often question passages that don't need editing,* such as an appropriate passive verb or a deliberate and emphatic use of repetition.

In the screen shot below the checker has flagged a direct repetition of *light* in the first sentence but left unflagged the other intrusive repetitions of the word. And the checker has flagged the entire second sentence because it is long, but in fact the sentence is grammatically correct and clear.

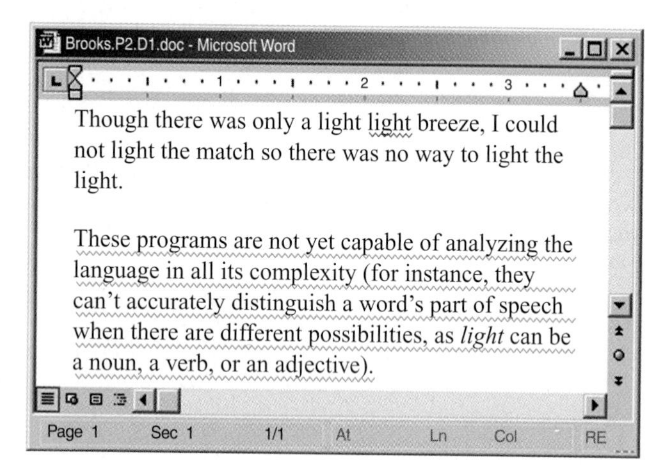

Each time a grammar and style checker questions something, you must determine whether a change is needed at all and what change will be most effective, and you must read your papers carefully on your own to find any errors the program missed.

PART I

Effective Sentences

CHECKLIST

Effective Sentences

Emphasis

☑ Make subjects and verbs of sentences focus on key actors and actions. (See opposite.)

☑ Stress main ideas by placing them first or last in a sentence. (See p. 12.)

☑ Link equally important ideas with coordination. (See p. 13.)

☑ De-emphasize less important ideas with subordination. (See p. 14.)

Conciseness

☑ Use the active voice to focus on key actors and actions. (See p. 15.)

☑ Cut empty words and unneeded repetition. (See pp. 15–16.)

☑ Avoid unneeded *there is* and *it is* constructions. (See p. 17.)

☑ Reduce word groups to their essence, and combine sentences where appropriate. (See pp. 16–17.)

Parallelism

☑ Use parallel constructions to show the equivalence of elements connected by *and, or, not only . . . but also,* and similar words. (See pp. 17–19.)

Variety and details

☑ Vary sentence lengths and structures to stress your main ideas and hold readers' attention. (See p. 20.)

☑ Provide details that make your sentences clear and interesting. (See p. 22.)

Appropriate words

☑ Use language appropriate for your writing situation. (See p. 22.)

☑ Avoid sexist and other biased language. (See p. 25.)

Exact words

☑ Choose words that are suited to your meaning and are concrete and specific. (See pp. 28–29.)

☑ Make words correct in idiom and also fresh, not clichéd. (See pp. 29–30.)

1 Emphasis

Emphatic writing leads readers to see your main ideas both within and among sentences. Besides the following strategies for achieving emphasis, see also the discussions of conciseness (p. 15) and variety (p. 19).

Note Many grammar and style checkers can spot some problems with emphasis, such as nouns made from verbs, passive voice, wordy phrases, and long sentences that may also be flabby and unemphatic. However, the checkers cannot help you identify the important ideas in your sentences or tell you whether those ideas receive appropriate emphasis for your meaning.

1a ▪ Subjects and verbs

The heart of every sentence is its subject,° which usually names the actor, and its verb,° which usually specifies the subject's action: *Children* [subject] *grow* [verb]. When these elements do not identify the key actor and action in the sentence, readers must find that information elsewhere and the sentence may be wordy and unemphatic. In the following sentences, the subjects and verbs are underlined:

Unemphatic — The intention of the company was to expand its workforce. A proposal was also made to diversify the backgrounds and abilities of employees.

These sentences are unemphatic because their key ideas (the company's intending and proposing) do not appear in their subjects and verbs. Revised, the sentences are not only clearer but more concise:

Revised — The company intended to expand its workforce. It also proposed to diversify the backgrounds and abilities of employees.

The constructions shown on the next page usually drain meaning from a sentence's subject and verb.

°The degree sign (°) marks every term defined in the Glossary of Terms, beginning on page 229.

Visit *www.ablongman.com/littlebrown* for added help and exercises on emphasis.

Nouns made from verbs

Nouns made from verbs can obscure the key actions of sentences and add words. These nouns include *intention* (from *intend*), *proposal* (from *propose*), *decision* (from *decide*), *expectation* (from *expect*), *persistence* (from *persist*), *argument* (from *argue*), and *inclusion* (from *include*).

Unemphatic — After the company made a decision to hire more disabled workers, its next step was the construction of wheelchair ramps and other facilities.

Revised — After the company decided to hire more disabled workers, it next constructed wheelchair ramps and other facilities.

Weak verbs

Weak verbs, such as *made* and *was* in the unemphatic sentence above, tend to stall sentences just where they should be moving and often bury key actions:

Unemphatic — The company is now the leader among businesses in complying with the 1990 Americans with Disabilities Act. Its officers make speeches on the act to business groups.

Revised — The company now leads other businesses in complying with the 1990 Americans with Disabilities Act. Its officers speak on the act to business groups.

Passive voice

Verbs in the passive voice° state actions received by, not performed by, their subjects. Thus the passive de-emphasizes the true actor of the sentence, sometimes omitting it entirely. Generally, prefer the active voice,° in which the subject performs the verb's action. (See p. 42 for more on passive and active voice.)

Unemphatic — The 1990 law is seen by most businesses as fair, but the costs of complying have sometimes been exaggerated.

Revised — Most businesses see the 1990 law as fair, but some opponents have exaggerated the costs of complying.

1b ▪ Sentence beginnings and endings

The beginning and ending of a sentence are the most emphatic positions, and the ending is usually more emphatic than the beginning. To emphasize information, place it first or last, reserving the middle for incidentals.

Unemphatic Education remains the single best means of economic advancement, despite its shortcomings.

Revised Education remains, despite its shortcomings, the single best means of economic advancement.

Generally, readers expect the beginning of a sentence to contain information that they already know or that you have already introduced. They then look to the ending for new information. In the unemphatic passage below, the second and third sentences both begin with new topics (underlined), while the old topics from the first sentence (the controversy and education) appear at the end:

Unemphatic Education almost means controversy these days, with rising costs and constant complaints about its inadequacies. But the value of schooling should not be obscured by the controversy. The single best means of economic advancement, despite its shortcomings, remains education.

In the revision below, the underlined old information begins each sentence and new information ends the sentence. The passage follows the pattern A→B. B→C. A→D.

Revised Education almost means controversy these days, with rising costs and constant complaints about its inadequacies. But the controversy should not obscure the value of schooling. Education remains, despite its shortcomings, the single best means of economic advancement.

1c ▪ Coordination

Use **coordination** to show that two or more elements in a sentence are equally important in meaning:

- Link two complete sentences (main clauses°) with a comma and a coordinating conjunction° (*and, but, or, nor, for, so, yet*).

  Independence Hall in Philadelphia is now restored, but fifty years ago it was in bad shape.

- Link two main clauses with a semicolon alone or a semicolon and a conjunctive adverb,° such as *however, indeed,* or *therefore.*

  The building was standing; however, it suffered from decay and vandalism.

- Within clauses, link words and word groups with a coordinating conjunction (*and, but, or, nor*) but no comma.

  The people and officials of the nation were indifferent to Independence Hall or took it for granted.

- Link main clauses, words, or phrases° with a correlative conjunction° such as *not only . . . but also* or *either . . . or*.

  People not only took the building for granted but also neglected it.

Notes A string of main clauses connected by *and* implies that all ideas are equally important and creates a dull, plodding rhythm. Use subordination (see below) to revise such excessive coordination. See also page 20.

Two punctuation errors, the comma splice and the fused sentence, can occur when you link main clauses. See pages 63–65.

1d ▪ Subordination

Use **subordination** to indicate that some elements in a sentence are less important than others for your meaning. Usually, the main idea appears in the main clause,° and supporting information appears in subordinate structures:

- Use a subordinate clause° beginning with a subordinating word such as *although, because, before, if, since, that, when, where, which,* or *who* (*whom*).

  Although production costs have declined, they are still high. [Stresses that costs are still high.]

  Costs, which include labor and facilities, are difficult to control. [Stresses that costs are difficult to control.]

- Use a phrase.°

  Despite some decline, production costs are still high.

  Costs, including labor and facilities, are difficult to control.

- Use a single word.

  Declining costs have not matched prices.
  Labor costs are difficult to control.

Generally, subordinate clauses give the most emphasis to secondary information, phrases give less, and single words give the least.

Note A subordinate clause or a phrase is not a complete sentence and should not be set off and punctuated as one. See pages 61–63 on sentence fragments.

2 Conciseness

Concise writing makes every word count. Conciseness is not the same as mere brevity: detail and originality should not be cut with needless words. Rather, the length of an expression should be appropriate to the thought.

Note Any grammar and style checker will identify at least some wordy structures, such as repeated words, weak verbs, passive voice, and *there is* and *it is* constructions. No checker can identify all wordy structures, however, nor can it tell you whether the structure is appropriate for your ideas.

2a ▪ Focusing on the subject and verb

Using the subjects° and verbs° of your sentences for the key actors and actions will tighten sentences. See page 12 under emphasis for a discussion of specific techniques:

- Avoiding nouns made from verbs, such as *intention* (from *intend*) and *decision* (from *decide*).
- Strengthening weak verbs, such as *is* and *make*.
- Rewriting the passive voice as active—for instance, changing *The star was seen by astronomers* to *Astronomers saw the star*.

2b ▪ Cutting empty words

Cutting words that contribute nothing to your meaning will make your writing move faster and work harder.

Wordy As far as I am concerned, because of the fact that a situation of discrimination continues in the field of medicine, women have not at the present time achieved equality with men.

Concise Because of continuing discrimination in medicine, women have not yet achieved equality with men.

The underlining in the wordy example above highlights the kinds of empty words described on the next page.

Visit *www.ablongman.com/littlebrown* for added help and exercises on conciseness.

- Some phrases add nothing to meaning:

all things considered	in a manner of speaking
a person by the name of	in my opinion
as far as I'm concerned	last but not least
for all intents and purposes	more or less

- Some abstract or general words, along with other words they require (such as *the* and *of*), pad sentences:

area	element	kind	situation
aspect	factor	manner	thing
case	field	nature	type

- Some word groups mean the same thing as single words:

For	Substitute
at all times	always
at the present time	now
at this point in time	now
for the purpose of	for
due to the fact that	because
because of the fact that	because
in the final analysis	finally

2c ▪ Cutting unneeded repetition

Repeating or restating key words from sentence to sentence can link the sentences and emphasize information the reader already knows (see p. 13). But unnecessary repetition weakens sentences and paragraphs.

Wordy Many unskilled workers without training in a particular job are unemployed and do not have any work.

Concise Many unskilled workers are unemployed.

Be especially alert to phrases that say the same thing twice. In the following examples, only the underlined words are needed:

circle around	repeat again
consensus of opinion	return again
cooperate together	square [round] in shape
the future to come	surrounding circumstances

2d ▪ Reducing clauses and phrases

Modifiers° can be expanded or contracted depending on the emphasis you want to achieve. (Generally, the longer a construction, the more emphasis it has.) When editing your sentences, consider whether any modifiers can be reduced without loss of emphasis or clarity.

Wordy The Channel Tunnel, which runs between Britain and France, bores through a bed of solid chalk that is twenty-three miles across.

Concise The Channel Tunnel between Britain and France bores through twenty-three miles of solid chalk.

2e ▪ Cutting *there is* or *it is*

Sentences beginning *there is* or *it is* (called expletive constructions°) are sometimes useful to emphasize a change in direction, but usually they just add needless words.

Wordy There were delays and cost overruns that plagued construction of the Channel Tunnel. It is the expectation of investors to earn profits at last, now that there are trains passing daily through the tunnel.

Concise Delays and cost overruns plagued construction of the Channel Tunnel. Investors expect to earn profits at last, now that trains pass daily through the tunnel.

2f ▪ Combining sentences

Often the information in two or more sentences can be combined into one tight sentence.

Wordy An unexpected problem with the Channel Tunnel is stowaways. The stowaways are mostly illegal immigrants. They are trying to smuggle themselves into England. They cling to train roofs and undercarriages.

Concise An unexpected problem with the Channel Tunnel is stowaways, mostly illegal immigrants who are trying to smuggle themselves into England by clinging to train roofs and undercarriages.

3 Parallelism

Parallelism matches the form of your sentence to its meaning: when your ideas are equally important, or par-

Visit *www.ablongman.com/littlebrown* for added help and exercises on parallelism.

allel, you express them in similar, or parallel, grammatical form.

> The air is dirtied by factories belching smoke and vehicles spewing exhaust.

Parallelism can also work like glue to link the sentences of a paragraph as well as the parts of a sentence:

> Pulleys are ancient machines for transferring power. Unfortunately, they are also inefficient machines.

Note A grammar and style checker cannot recognize faulty parallelism because it cannot recognize the relations among ideas. You will need to find and revise problems with parallelism on your own.

3a ▪ Parallelism with *and, but, or, nor, yet*

The coordinating conjunctions° *and, but, or, nor,* and *yet* signal a need for parallelism.

> The industrial base was shifting and shrinking.
>
> Politicians seldom acknowledged the problem or proposed alternatives.
>
> Industrial workers were understandably disturbed that they were losing their jobs and that no one seemed to care.

Nonparallel: The reasons why steel companies kept losing money were that their plants were inefficient, high labor costs, and foreign competition was increasing.

Revised: The reasons steel companies kept losing money were inefficient plants, high labor costs, and increasing foreign competition.

Notes Parallel elements match in structure, but they need not match word for word. In the preceding example, each element consists of at least one modifier° and a noun,° but two of the elements also include an additional modifier.

Be careful not to omit needed words in parallel structures.

Nonparallel: Given training, workers can acquire the skills and interest in other jobs. [Idiom dictates different prepositions with *skills* and *interest*.]

Revised: Given training, workers can acquire the skills for and interest in other jobs.

3b ▪ Parallelism with *both . . . and, either . . . or,* and so on

Correlative conjunctions° stress equality and balance between elements. The correlative conjunctions include *both . . . and, either . . . or, neither . . . nor, not only . . . but also,* and *whether . . . or.* Parallelism confirms the equality between elements: the words after the first and second connectors must match.

Nonparallel Huck Finn learns not only that human beings have an enormous capacity for folly but also enormous dignity. [The first element includes *that human beings have;* the second element does not.]

Revised Huck Finn learns that human beings have not only an enormous capacity for folly but also enormous dignity. [Repositioning *not only* makes the two elements parallel.]

3c ▪ Parallelism with lists, headings, and outlines

The items in a list or outline should be parallel. Parallelism is essential in the headings that divide a paper into sections (see pp. 201–02).

Nonparallel	Revised
Changes in Renaissance England	Changes in Renaissance England
1. Extension of trade routes	1. Extension of trade routes
2. Merchant class became more powerful	2. Increased power of the merchant class
3. The death of feudalism	3. Death of feudalism
4. Upsurging of the arts	4. Upsurge of the arts
5. Religious quarrels began	5. Rise of religious quarrels

4 Variety and Details

To make your writing interesting as well as clear, use varied sentences that are well textured with details.

Visit *www.ablongman.com/littlebrown* for added help with variety and details.

Note Some grammar and style checkers will flag long sentences, and you can check for appropriate variety in a series of such sentences. But generally these programs cannot help you see where variety may be needed because they cannot recognize the relative importance and complexity of your ideas. Nor can they suggest where you should add details. To edit for variety and detail, you need to listen to your sentences and determine whether they clarify your meaning.

4a ▪ Varied sentence lengths and structures

In most contemporary writing, sentences tend to vary from about ten to about forty words, with an average of fifteen to twenty-five words. If your sentences are all at one extreme or the other, your readers may have difficulty locating main ideas and seeing the relations among them.

- Break a sequence of long sentences into shorter, simpler ones that stress key ideas.
- Combine a sequence of short sentences with coordination (p. 13) and subordination (p. 14) to show relationships and stress main ideas.

A good way to focus and hold readers' attention is to vary the structure of sentences so that they do not all follow the same pattern, like soldiers in a parade.

Varied sentence structures

A long sequence of main clauses can make all ideas seem equally important and create a plodding rhythm, as in the unvaried passage below. You want to emphasize your key subjects and verbs, moving in each sentence from old information to new (see p. 13). Subordinating less important information (bracketed in the revised passage) can help you achieve this emphasis.

Unvaried

The moon is now drifting away from the earth. It moves away about one inch a year. This movement is lengthening our days, and they increase about a thousandth of a second every century. Forty-seven of our present days will someday make up a month. We might eventually lose the moon altogether. Such great planetary movement rightly concerns astronomers, but it need not worry us. It will take 50 million years.

Revised

The moon is now drifting away from the earth [about one inch a year.] [At a thousandth of a second every century,] this movement is lengthening our days. Forty-seven of our present days will someday make up a month, [if we don't eventually lose the moon altogether.] Such great planetary movement rightly concerns astronomers, but it need not worry us. It will take 50 million years.

Varied sentence beginnings

An English sentence often begins with its subject, which generally captures old information from a preceding sentence (see p. 13):

The defendant's lawyer was determined to break the prosecution's witness. He relentlessly cross-examined the stubborn witness for a week.

However, an unbroken sequence of sentences beginning with the subject quickly becomes monotonous:

Monotonous

The defendant's lawyer was determined to break the prosecution's witness. He relentlessly cross-examined the witness for a week. The witness had expected to be dismissed in an hour and was visibly irritated. She did not cooperate. She was reprimanded by the judge.

Revised

The defendant's lawyer was determined to break the prosecution's witness. For a week he relentlessly cross-examined the witness. Expecting to be dismissed in an hour, the witness was visibly irritated. She did not cooperate. Indeed, she was reprimanded by the judge.

The underlined expressions represent the most common choices for varying sentence beginnings:

- Adverb modifiers, such as *For a week* (modifies the verb *cross-examined*).
- Adjective modifiers, such as *Expecting to be dismissed in an hour* (modifies *witness*)
- Transitional expressions, such as *Indeed*. (See *transitional expression*, p. 239, for a list.)

Varied word order

Occasionally, to achieve special emphasis, reverse the usual word order of a sentence.

A dozen witnesses testified, and the defense attorney barely questioned eleven of them. The twelfth, however, he grilled. [Compare normal word order: *He grilled the twelfth, however.*]

4b ▪ Details

Relevant details such as facts and examples create the texture and life that keep readers alert and help them grasp your meaning. For instance:

Flat

Constructed after World War II, Levittown, New York, comprised thousands of houses in two basic styles. Over the decades, residents have altered the houses so dramatically that the original styles are often unrecognizable.

Detailed

Constructed on potato fields after World War II, Levittown, New York, comprised more than 17,000 houses in Cape Cod and ranch styles. Over the decades, residents have added expansive columned porches, punched dormer windows through roofs, converted garages to sun porches, and otherwise altered the houses so dramatically that the original styles are often unrecognizable.

5 Appropriate Words

American academic writing relies on a dialect called **standard American English.** The dialect is also used in business, the professions, government, the media, and other sites of social and economic power where people of diverse backgrounds must communicate with one another. It is "standard" not because it is better than other forms of English, but because it is accepted as the common language, much as the dollar bill is accepted as the common currency.

The vocabulary of standard American English is huge, allowing expression of an infinite range of ideas and feelings; but it does exclude words that only some groups of people use, understand, or find inoffensive. Some of these more limited vocabularies should be avoided altogether; others should be used cautiously and in relevant situations, as when aiming for a special effect with an audience you know will appreciate it.

Note Many grammar and style checkers can be set to flag potentially inappropriate words, such as nonstan-

Visit *www.ablongman.com/littlebrown* for added help and exercises on appropriate words.

dard language, slang, colloquialisms, and gender-specific terms (*manmade, mailman*). However, the checker can flag only words listed in its dictionary of questionable words. For example, a checker flagged *businessman* as potentially sexist in *A successful businessman puts clients first,* but the checker did not flag *his* in *A successful businessperson listens to his clients*. If you use a checker to review your language, you'll need to determine whether a flagged word is or is not appropriate for your writing situation, and you'll still need to hunt for possibly inappropriate words on your own.

5a ▪ Nonstandard dialect

Like many countries, the United States includes scores of regional, social, or ethnic groups with their own distinct **dialects,** or versions of English. Standard American English is one of those dialects, and so are Black English, Appalachian English, and Creole. All the dialects of English share many features, but each also has its own vocabulary, pronunciation, and grammar.

If you speak a dialect of English besides standard American English, be careful about using your dialect in situations where standard English is the norm, such as in academic or business writing. Dialects are not wrong in themselves, but forms imported from one dialect into another may still be perceived as unclear or incorrect. When you know standard English is expected in your writing, edit to eliminate expressions in your dialect that you know (or have been told) differ from standard English. These expressions may include *theirselves, hisn, them books*, and others labeled "nonstandard" by a dictionary. They may also include certain verb forms, as discussed on pages 33 and 34.

Your participation in the community of standard American English does not require you to abandon your own dialect. You may want to use it in writing you do for yourself, such as journals, notes, and drafts, which should be composed as freely as possible. You may want to quote it in an academic paper, as when analyzing or reporting conversation in dialect. And, of course, you will want to use it with others who speak it.

5b ▪ Slang

Slang is the insider language used by a group, such as musicians or football players, to reflect common experiences and to make technical references efficient. The following example is from an essay on the slang of "skaters" (skateboarders):

> Curtis slashed ultra-punk crunchers on his longboard, while the Rube-man flailed his usual Gumbyness on tweaked frontsides and lofty fakie ollies.
>
> —Miles Orkin, "Mucho Slingage by the Pool"

Though valuable within a group, slang is often too private or imprecise for academic or business writing.

5c ▪ Colloquial language

Colloquial language is the everyday spoken language, including expressions such as *go crazy, get along with, a lot, kids* (for *children*), and *stuff* (for possessions or other objects). Dictionaries label this language "informal" or "colloquial."

Colloquial language suits informal writing, and an occasional colloquial word can help you achieve a desired emphasis in otherwise formal writing. But most colloquial language is not precise enough for academic or career writing.

5d ▪ Technical words

All disciplines and professions rely on specialized language that allows the members to communicate precisely and efficiently with each other. Chemists, for instance, have their *phosphatides,* and literary critics have their *subtexts.* Use the terms of a discipline or profession when you are writing within it.

However, when you are writing for a nonspecialist audience, avoid unnecessary technical terms and carefully define the terms you must use.

5e ▪ Indirect and pretentious writing

Small, plain, and direct words are usually preferable to big, showy, or evasive words. Take special care to avoid the following:

- **Euphemisms** are presumably inoffensive words that substitute for words deemed potentially offensive or too blunt, such as *passed away* for *died* or *misspeak* for *lie.* Use euphemisms only when you know that blunt, truthful words would needlessly hurt or offend members of your audience.
- **Double talk** (at times called **doublespeak** or **weasel words**) is language intended to confuse or to be misunderstood: the *revenue enhancement* that is really a

tax, the *biodegradable* bags that still last decades. Double talk has no place in honest writing.

- **Pretentious writing** is fancy language that is more elaborate than its subject requires. Choose your words for their exactness and economy. The big, ornate word may be tempting, but pass it up. Your readers will be grateful.

Pretentious — To perpetuate our endeavor of providing funds for our elderly citizens as we do at the present moment, we will face the exigency of enhanced contributions from all our citizens.

Revised — We cannot continue to fund Social Security and Medicare for the elderly unless we raise taxes.

5f ▪ Sexist and other biased language

Even when we do not mean it to, our language can reflect and perpetuate hurtful prejudices toward groups of people. Such biased language can be obvious—words such as *nigger, honky, mick, kike, fag, dyke,* or *broad.* But it can also be subtle, generalizing about groups in ways that may be familiar but that are also inaccurate or unfair.

Biased language reflects poorly on the user, not on the person or persons whom it mischaracterizes or insults. Unbiased language does not submit to false generalizations. It treats people respectfully as individuals and labels groups as they wish to be labeled.

Stereotypes of race, ethnicity, and other characteristics

A **stereotype** characterizes and judges people simply on the basis of their membership in a group: *Men are uncommunicative. Women are emotional. Liberals want to raise taxes. Conservatives are affluent.*

In your writing, avoid statements about the traits of whole groups that may be true of only some members. Be especially cautious about substituting such statements for the evidence you should be providing instead.

Stereotype — Elderly drivers should have their licenses limited to daytime driving only. [Asserts that all elderly people are poor night drivers.]

Revised — Drivers with impaired night vision should have their licenses limited to daytime driving only.

Some stereotypes have become part of the language, but they are still potentially offensive.

Stereotype	The administrators are too blind to see the need for a new gymnasium.
Revised	The administrators do not understand the need for a new gymnasium.

Sexist language

Sexist language distinguishes needlessly between men and women in such matters as occupation, ability, behavior, temperament, and maturity. It can wound or irritate readers and indicates the writer's thoughtlessness or unfairness. The following guidelines can help you eliminate sexist language from your writing.

- Avoid demeaning and patronizing language—for instance, identifying women and men differently or trivializing either gender.

Sexist	Dr. Keith Kim and Lydia Hawkins coauthored the article.
Revised	Dr. Keith Kim and Dr. Lydia Hawkins coauthored the article.
Revised	Keith Kim and Lydia Hawkins coauthored the article.
Sexist	Ladies are entering almost every occupation formerly filled by men.
Revised	Women are entering almost every occupation formerly filled by men.

- Avoid occupational or social stereotypes that assume a role or profession is exclusively male or female.

Sexist	The considerate doctor commends a nurse when she provides his patients with good care.
Revised	The considerate doctor commends a nurse who provides good care for patients.

- Avoid using *man* or words containing *man* to refer to all human beings. Some alternatives:

businessman	businessperson
chairman	chair, chairperson
congressman	representative in Congress, legislator
craftsman	craftsperson, artisan
layman	layperson
mankind	humankind, humanity, human beings, people

manpower	personnel, human resources
policeman	police officer
salesman	salesperson, sales representative

Sexist Man has not reached the limits of social justice.

Revised Humankind [or Humanity] has not reached the limits of social justice.

Sexist The furniture consists of manmade materials.

Revised The furniture consists of synthetic materials.

- Avoid using *he* to refer to both genders. (See also p. 50.)

Sexist The newborn child explores his world.

Revised Newborn children explore their world. [Use the plural for the pronoun and the word it refers to.]

Revised The newborn child explores the world. [Avoid the pronoun altogether.]

Revised The newborn child explores his or her world. [Substitute male and female pronouns.]

Use the last option sparingly—only once in a group of sentences and only to stress the singular individual.

Inappropriate labels

Labels for groups of people can be shorthand stereotypes and can be discourteous when they ignore readers' preferences. Although sometimes dismissed as "political correctness," sensitivity in applying labels hurts no one and helps gain your readers' trust and respect.

- Avoid labels that (intentionally or not) disparage the person or group you refer to. A person with emotional problems is not a *mental patient.* A person with cancer is not a *cancer victim.* A person using a wheelchair is not *wheelchair-bound.*
- Use names for racial, ethnic, and other groups that reflect the preferences of each group's members, or at least many of them. Examples of current preferences include *African American* or *black, latino/latina* (for Americans of Spanish-speaking descent), and *people with disabilities* (rather than *the handicapped*). But labels change often. To learn how a group's members wish to be labeled, ask them directly, attend to usage in reputable periodicals, or check a recent dictionary.

6 Exact Words

6a

To write clearly and effectively, you will want to find the words that fit your meaning exactly and convey your attitude precisely.

Note A grammar and style checker can provide some help with inexact language. For instance, you can set it to flag commonly confused words (such as *continuous/continual*), misused prepositions in idioms (such as *accuse for* instead of *accuse of*), and clichés. But the checker can flag only words stored in its dictionary. It can't help at all with other problems discussed in this section. You'll need to read your work carefully on your own.

6a • The right word for your meaning

One key to helping readers understand you is to use words according to their established meanings.

- Become acquainted with a dictionary. Consult it whenever you are unsure of a word's meaning.
- Distinguish between similar-sounding words that have widely different meanings.

  Inexact Older people often suffer infirmaries [places for the sick].

  Exact Older people often suffer infirmities [disabilities].

  Some words, called **homonyms,** sound exactly alike but differ in meaning: for example, *principal/principle* or *rain/reign/rein*. (Many homonyms and near-homonyms are listed in the Glossary of Usage, p. 219.)
- Distinguish between words with related but distinct meanings.

  Inexact Television commercials continuously [unceasingly] interrupt programming.

  Exact Television commercials continually [regularly] interrupt programming.

Visit ***www.ablongman.com/littlebrown*** for added help and exercises on exact words.

- Distinguish between words that have similar basic meanings but different emotional associations, or **connotations**.

  It is a daring plan. [The plan is bold and courageous.]
  It is a reckless plan. [The plan is thoughtless and risky.]

Many dictionaries list and distinguish such **synonyms**, words with approximately, but often not exactly, the same meanings.

6b ▪ Concrete and specific words

Clear, exact writing balances abstract and general words, which outline ideas and objects, with concrete and specific words, which sharpen and solidify.

- **Abstract words** name qualities and ideas: *beautiful, management, culture, freedom, awesome.* **Concrete words** name things we can know by our five senses of sight, hearing, touch, taste, and smell: *sleek, humming, brick, bitter, musty.*
- **General words** name classes or groups of things, such as *buildings, weather,* or *birds,* and include all the varieties of the class. **Specific words** limit a general class, such as *buildings,* by naming one of its varieties, such as *skyscraper, Victorian courthouse,* or *hut.*

Abstract and general statements need development with concrete and specific details. For example:

Vague The size of his hands made his smallness real. [How big were his hands? How small was he?]

Exact Not until I saw his delicate, doll-like hands did I realize that he stood a full head shorter than most other men.

6c ▪ Idioms

Idioms are expressions in any language that do not fit the rules for meaning or grammar—for instance, *put up with, plug away at, make off with.*

Because they are not governed by rules, idioms usually cause particular difficulty for people learning to speak and write a new language. But even native speakers of English misuse some idioms involving prepositions,° such as *agree on a plan, agree to a proposal,* and *agree with a person* or *occupied by a person, occupied in study,* and *occupied with a thing.*

When in doubt about an idiom, consult your dictionary under the main word (*agree* and *occupy* in the examples). (See also pp. 37–38 on verbs with particles and p. 132 on signal phrases.)

6d ▪ Clichés

Clichés, or **trite expressions**, are phrases so old and so often repeated that they have become stale. Examples include *better late than never, beyond the shadow of a doubt, face the music, green with envy, ladder of success, point with pride, sneaking suspicion,* and *wise as an owl.*

Clichés may slide into your drafts. In editing, be wary of any expression you have heard or used before. Substitute fresh words of your own, or restate the idea in plain language.

PART II

Grammatical Sentences

VERBS

PRONOUNS

MODIFIERS

SENTENCE FAULTS

CHECKLIST

Grammatical Sentences

This checklist focuses on the most common and potentially confusing grammatical errors. See the contents inside the back cover for a more detailed guide to this part.

Verbs

☑ Use the correct forms of irregular verbs such as *has broken* [not *has broke*]. (See opposite.)

☑ Use helping verbs where required, as in *she has been* [not *she been*]. (See p. 34.)

☑ Match verbs to their subjects, as in *The list of items is* [not *are*] *long.* (See p. 43.)

Pronouns

☑ Match pronouns to the words they refer to, as in *Each of the women had her* [not *their*] *say.* (See p. 50.)

☑ Make pronouns refer clearly to the words they substitute for, avoiding uncertainties such as *Jill thanked Tracy when she* [Jill or Tracy?] *arrived.* (See p. 51.)

☑ Use pronouns consistently, avoiding shifts such as *When one enters college, you meet new ideas.* (See p. 53.)

Modifiers

☑ Place modifiers close to the words they describe, as in *Trash cans without lids invite animals* [not *Trash cans invite animals without lids*]. (See p. 58.)

☑ Make each modifier clearly modify another word in the sentence, as in *Jogging, she pulled a muscle* [not *Jogging, a muscle was pulled*]. (See p. 60.)

Sentence faults

☑ Make every sentence complete, with its own subject and verb, and be sure none is a freestanding subordinate clause—for instance, *But first she called the police* [not *But first called the police*]; *New stores open weekly* [not *New stores weekly*]; and *The new cow calved after the others did* [not *The new cow calved. After the others did*]. (See p. 61.)

☑ Within sentences, link main clauses with a comma and a coordinating conjunction (*Cars jam the roadways, and they contribute to smog*), with a semicolon (*Many parents did not attend; they did not want to get involved*), or with a semicolon and a conjunctive adverb (*The snow fell heavily; however, it soon melted*). (See p. 63.)

7 Verb Forms

Verb forms may give you trouble when the verb is irregular, when you omit certain endings, or when you need to use helping verbs.

Note A grammar and style checker may flag incorrect verb forms, but it may also fail to do so. For example, a checker flagged *The runner stealed second base* (*stole* is correct) but not *The runner had steal second base* (*stolen* is correct). When in doubt about verb forms, consult a dictionary or the links at this book's Web site.

7a ▪ *Sing/sang/sung* and other irregular verbs

Most verbs are **regular:** their past-tense form° and past participle° end in *-d* or *-ed*:

> Today the birds migrate. [Plain form° of verb.]
>
> Yesterday the birds migrated. [Past-tense form.]
>
> In the past the birds have migrated. [Past participle.]

About two hundred **irregular verbs** in English create their past-tense form and past participle in some way besides adding *-d* or *-ed*.

> Today the birds fly. They begin migration. [Plain form.]
>
> Yesterday the birds flew. They began migration. [Past-tense form.]
>
> In the past the birds have flown. They have begun migration. [Past participle.]

You can find a verb's forms by looking up the plain form in a dictionary. For a regular verb, the dictionary will give the *-d* or *-ed* form. For an irregular verb, the dictionary will give the past-tense form and then the past participle. If the dictionary gives only one irregular form after the plain form, then the past-tense form and past participle are the same (*think, thought, thought*).

Visit *www.ablongman.com/littlebrown* for added help and exercises on verb forms.

7b • Helping verbs

Helping verbs combine with some verb forms to indicate time and other kinds of meanings, as in *can run, might suppose, will open, was sleeping, had been eaten.* The main verb° in these phrases is the one that carries the main meaning (*run, suppose, open, sleeping, eaten*).

Required helping verbs

Standard American English requires helping verbs in certain situations:

- The main verb ends in *-ing:*

  Archaeologists are conducting fieldwork all over the world. [Not *Archaeologists conducting. . . .*]

- The main verb is *been* or *be:*

  Many have been fortunate in their discoveries. [Not *Many been. . . .*]

  Some could be real-life Indiana Joneses. [Not *Some be. . . .*]

- The main verb is a past participle,° such as *talked, begun,* or *thrown:*

  The researchers have given interviews on TV. [Not *The researchers given. . . .*]

In these examples, omitting the helping verb would create an incomplete sentence, or **sentence fragment** (p. 61).

Combinations of helping and main verbs ESL

Helping verbs and main verbs combine in specific ways.

Note The main verb in a verb phrase (the one carrying the main meaning) does not change to show a change in subject or time: *she has sung, you had sung.* Only the helping verb may change, as in these examples.

Form of *be* + present participle

Create the progressive tenses° with *be, am, is, are, was, were,* or *been* followed by the main verb's present participle° (ending in *-ing*).

Faulty She is work on a new book.
Revised She is working on a new book.

Faulty She has been work on it several months.
Revised She has been working on it several months.

Note Verbs that express mental states or activities rather than physical actions do not usually appear in the progressive tenses. These verbs include *adore, appear, believe, belong, have, hear, know, like, love, need, see, taste, think, understand,* and *want.*

Faulty She is wanting to understand contemporary ethics.

Revised She wants to understand contemporary ethics.

Form of *be* + past participle

Create the passive voice° with *be, am, is, are, was, were, being,* or *been* followed by the main verb's past participle° (usually ending in *-d* or *-ed* or, for irregular verbs, in *-t* or *-n*).

Faulty Her last book was complete in four months.

Revised Her last book was completed in four months.

Faulty It was bring to the President's attention.

Revised It was brought to the President's attention.

Note Only transitive verbs° may form the passive voice.

Faulty A philosophy conference was occurred that week. [*Occur* is not a transitive verb.]

Revised A philosophy conference occurred that week.

Form of *have* + past participle

To create one of the perfect tenses,° use the main verb's past participle preceded by a form of *have,* such as *has, had, have been,* or *will have had.*

Faulty Some students have complain about the lab.

Revised Some students have complained about the lab.

Faulty Money has not been spend on the lab in years.

Revised Money has not been spent on the lab in years.

Form of *do* + plain form

Always with the plain form° of the main verb, three forms of *do* serve as helping verbs: *do, does, did.*

Faulty Safety concerns do exists.

Revised Safety concerns do exist.

Faulty Didn't the lab closed briefly last year?

Revised Didn't the lab close briefly last year?

Modal + plain form

Most **modal** helping verbs combine with the plain form of the main verb to convey ability, possibility, necessity, and other meanings. The modals include *be able to,* *be supposed to, can, could, had better, have to, may, might, must, ought to, shall, should, used to, will,* and *would.*

Faulty The lab equipment may causes injury.
Revised The lab equipment may cause injury.

Faulty The school ought to replaced it.
Revised The school ought to replace it.

Note When a modal combines with another helping verb, the main verb generally changes from the plain form to a past participle:

Faulty The equipment could have fail.
Revised The equipment could have failed.

7c ▪ Verb + gerund or infinitive ESL

A **gerund** is the *-ing* form of a verb used as a noun (*Smoking kills*). An **infinitive** is the plain form° of the verb plus *to* (*Try to quit*). Gerunds and infinitives may follow certain verbs but not others. And sometimes the use of a gerund or infinitive with the same verb changes the meaning of the verb.

Either gerund or infinitive

A gerund or an infinitive may follow these verbs with no significant difference in meaning: *begin, can't bear, can't stand, continue, hate, hesitate, like, love, prefer, start.*

The pump began working. The pump began to work.

Meaning change with gerund or infinitive

With four verbs—*forget, remember, stop,* and *try*—a gerund has quite a different meaning from an infinitive.

The engineer stopped watching the pump. [She no longer watched.]

The engineer stopped to watch the pump. [She stopped in order to watch.]

Gerund, not infinitive

Do not use an infinitive after these verbs: *admit, adore, appreciate, avoid, consider, deny, detest, discuss, dislike, enjoy, escape, finish, imagine, keep, mind, miss, practice, put off, quit, recall, resent, resist, risk, suggest, tolerate, understand.*

Faulty She suggested to check the pump.
Revised She suggested checking the pump.

Infinitive, not gerund

Do not use a gerund after these verbs: *agree, ask, assent, beg, claim, decide, expect, have, hope, manage, mean, offer, plan, pretend, promise, refuse, say, wait, want, wish.*

Faulty She decided checking the pump.

Revised She decided to check the pump.

Noun or pronoun + infinitive

Some verbs may be followed by an infinitive alone or by a noun° or pronoun° and an infinitive: *ask, beg, choose, dare, expect, help, need, promise, want, wish, would like.* A noun or pronoun changes the meaning.

She expected to watch.

She expected her workers to watch.

Some verbs *must* be followed by a noun or pronoun before an infinitive: *advise, allow, cause, challenge, command, convince, encourage, forbid, force, hire, instruct, order, permit, persuade, remind, require, teach, tell, warn.*

She instructed her workers to watch.

Do not use *to* before the infinitive when it comes after one of the following verbs and a noun or pronoun: *feel, have, hear, let, make* ("force"), *see, watch.*

She let her workers learn by observation.

7d ▪ Verb + particle ESL

Some verbs consist of two words: the verb itself and a **particle,** a preposition° or adverb° that affects the meaning of the verb, as in *Look up the answer* (research the answer) or *Look over the answer* (check the answer). Many of these two-word verbs, also called idioms, are defined in dictionaries. (For more on idioms, see pp. 29–30.)

Some two-word verbs may be separated in a sentence; others may not.

Inseparable two-word verbs

Verbs and particles that may not be separated by any other words include the following: *catch on, get along, give in, go out, grow up, keep on, look into, run into, run out of, speak up, stay away, take care of.*

Faulty Children grow quickly up.

Revised Children grow up quickly.

Separable two-word verbs

Most two-word verbs that take direct objects° may be separated by the object.

> Parents help out their children.
>
> Parents help their children out.

If the direct object is a pronoun,° the pronoun *must* separate the verb from the particle.

> Faulty Parents help out them.
>
> Revised Parents help them out.

The separable two-word verbs include the following: *call off, call up, fill out, fill up, give away, give back, hand in, help out, look over, look up, pick up, point out, put away, put back, put off, take out, take over, try on, try out, turn down.*

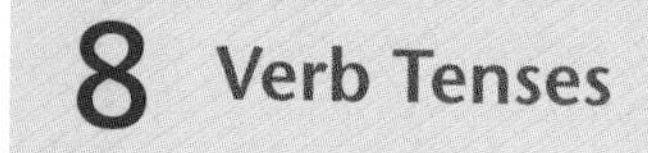

8 Verb Tenses

The **tense** of a verb expresses the time its action occurred. Definitions and examples of the verb tenses appear on pages 238–39. The following are the most common trouble spots.

Note Grammar and style checkers can provide little help with incorrect verb tenses and tense sequences because correctness usually depends on meaning. Proofread carefully yourself to catch errors in tense or tense sequence.

8a ▪ Uses of the present tense (*sing*)

Most academic and business writing uses the past tense° (*the rebellion occurred*), but the present tense has several distinctive uses:

> Action occurring now
> We define the problem differently.
>
> Habitual or recurring action
> Banks regularly undergo audits.
>
> A general truth
> The earth is round.

Visit ***www.ablongman.com/littlebrown*** for added help and exercises on verb tenses.

Discussion of literature, film, and so on

Huckleberry Finn has adventures we all envy.

Future time

Funding ends in less than a year.

8b ▪ Uses of the perfect tenses (*have/had/will have sung*)

The perfect tenses° generally indicate an action completed before another specific time or action. The present perfect tense° also indicates action begun in the past and continued into the present.

present perfect
The dancer has performed here only once.

past perfect
The dancer had trained in Asia before his performance here ten years ago.

future perfect
He will have performed here again by next month.

8c ▪ Consistency in tense

Within a sentence, the tenses of verbs and verb forms need not be identical as long as they reflect actual changes in time: *Ramon will graduate from college twenty years after his father arrived in America.* But needless shifts in tense will confuse or distract readers.

Inconsistent Immediately after Booth shot Lincoln, Major Rathbone threw himself upon the assassin. But Booth pulls a knife and plunges it into the major's arm.

Revised Immediately after Booth shot Lincoln, Major Rathbone threw himself upon the assassin. But Booth pulled a knife and plunged it into the major's arm.

8d ▪ Sequence of tenses

When the tenses in a sentence are in **sequence,** the verbs in the main clause° and the subordinate clause° relate appropriately for meaning. Problems with tense sequence often occur with the past tense and past perfect tense.°

Faulty When researchers tried (past) to review the study, many of the original participants died (past).

Revised When researchers tried (past) to review the study, many of the original participants had died (past perfect). [The deaths had occurred before the review.]

Faulty — Because other participants refused (past) interviews, the review had been terminated (past perfect).

Revised — Because other participants refused (past) interviews, the review was terminated (past). [The refusal occurred before the termination.]

Other tense-sequence problems occur with the distinctive verb forms of **conditional sentences,** in which a subordinate clause begins *if, when,* or *unless* and the main clause states the result.

Faulty — If voters have (present) more confidence, they would vote (*would* + verb) more often.

Revised — If voters had (past) more confidence, they would vote (*would* + verb) more often.

See below for more on verbs in conditional sentences.

9 Verb Mood

The **mood** of a verb indicates whether a sentence is a statement or a question (*The theater needs help. Can you help the theater?*), a command (*Help the theater*), or a suggestion, desire, or other nonfactual expression (*I wish I were an actor*).

Note A grammar and style checker may spot some simple errors in mood, but it may miss others. For example, a checker flagged *I wish I was home* (should be *were*) but not *If I had a hammer, I will hammer in the morning* (should be *would*).

9a ▪ Subjunctive mood: *I wish I were*

The **subjunctive mood** expresses a suggestion, requirement, or desire, or it states a condition that is contrary to fact (that is, imaginary or hypothetical).

- Suggestion or requirement with the verb *ask, insist, urge, require, recommend,* or *suggest:* use the verb's plain form° with all subjects.

Visit ***www.ablongman.com/littlebrown*** for added help and exercises on verb mood.

Rules require that every donation be mailed.

- Desire or present condition contrary to fact: use the verb's past-tense form;° for *be,* use the past-tense form *were.*

If the theater were in better shape and had more money, its future would be guaranteed.

I wish I were able to donate money.

- Past condition contrary to fact: use the verb's past perfect form° (*had* + past participle).

The theater would be better funded if it had been better managed.

Note In a sentence expressing a condition contrary to fact, (1) use *have,* not *of,* after *would* or *could* in the main clause; and (2) do not use *would* or *could* in the subordinate clause:

Faulty	People would of helped if they would have known.
Revised	People would have helped if they had known.

9b • Consistency in mood

Shifts in mood within a sentence or among related sentences can be confusing. Such shifts occur most frequently in directions.

Inconsistent	Dissolve the crystals in the liquid. Then you should heat the solution to 120°C. [The first sentence is a command, the second a statement.]
Revised	Dissolve the crystals in the liquid. Then heat the solution to 120°C. [Consistent commands.]

10 Verb Voice

The **voice** of a verb tells whether the subject° of the sentence performs the action (**active voice**) or is acted upon (**passive voice**).

Visit *www.ablongman.com/littlebrown* for added help and exercises on verb voice.

Active voice	Commercial services expand participation on the Internet.
Passive voice	Participation on the Internet is expanded by commercial services.

10a ▪ Active voice vs. passive voice

The active voice always names the actor in a sentence (whoever performs the verb's action), whereas the passive voice puts the actor in a phrase after the verb or even omits the actor altogether. Thus the active voice is usually more clear, emphatic, and concise than the passive voice.

Weak passive	The Internet is used for research by scholars, and its expansion to the general public was criticized by some.
Strong active	Scholars use the Internet for research, and some criticized its expansion to the general public.

The passive voice is useful in two situations: when the actor is unknown and when the actor is unimportant or less important than the object of the action.

> The Internet was established in 1969 by the Department of Defense. The network has now been extended both nationally and internationally. [In the first sentence the writer wishes to stress the Internet. In the second sentence the actor is unknown or too complicated to name.]

> After the solution had been cooled to 10°C, the acid was added. [The person who cooled and added, perhaps the writer, is less important than the actions. Passive sentences are common in scientific writing.]

Note Most grammar and style checkers can be set to spot the passive voice. But they will then flag every instance, both appropriate and ineffective. And they will flag as passive some phrases that are actually a form of *be* plus an adjective, such as *We were delighted*. You'll need to decide whether flagged phrases require revision.

10b ▪ Consistency in voice

A shift in voice (and subject) within or between sentences can be awkward or even confusing.

Inconsistent	Commercial services provide fairly inexpensive Internet access, and navigation is made easy by them.
Revised	Commercial services provide fairly inexpensive Internet access, and they make navigation easy.

11 Agreement of Subject and Verb

A subject° and its verb° should agree in number°—singular with singular, plural with plural.

> Many Japanese Americans (subject) live (verb) in Hawaii and California.

> Daniel Inouye (subject) was (verb) the first Japanese American in Congress.

Note Grammar and style checkers will catch most simple errors in subject-verb agreement, such as *Margaret and John is late,* but they are less reliable with more complicated sentences. For example, a checker flagged *Is Tom and Cathy going with us?* (should be *Are*) but not *Are Tom or Cathy going with us?* (should be *Is*). Checkers may also mistakenly flag correct agreement and then suggest "corrections" that are wrong. Do not automatically accept a checker's pointers, and proofread your work carefully on your own.

11a

11a ▪ *-s* ending for noun *or* verb, but not both

An *-s* or *-es* ending does opposite things to nouns and verbs: it usually makes a noun *plural,* but it always makes a present-tense verb *singular*. Thus if the subject noun is plural, it will probably end in *-s* or *-es* and the verb will not. If the subject is singular, it will not end in *-s* and the verb will.

Singular noun	Plural noun
The boy plays.	The boys play.
The bird soars.	The birds soar.
The street is busy.	The streets are busy.
The town has a traffic problem.	The towns have a traffic problem.
The new light does not [or doesn't] help.	The new lights do not [or don't] help.1

ESL Most noncount nouns—those that do not form plurals—take singular verbs: *That information is helpful.* (See pp. 57–58 for more on noncount nouns.)

Visit ***www.ablongman.com/littlebrown*** for added help and exercises on subject-verb agreement.

11b ▪ Words between subject and verb

A catalog of courses and requirements often baffles [not baffle] students.

The requirements stated in the catalog are [not is] unclear.

Phrases beginning with *as well as, together with, along with,* and *in addition to* do not change the number of the subject.

The president, as well as the deans, has [not have] agreed to revise the catalog.

11c ▪ Subjects with *and*

Frost and Roethke were American poets who died in the same year.

Note When *each* or *every* precedes the compound subject, the verb is usually singular.

Each man, woman, and child has a right to be heard.

11d ▪ Subjects with *or* or *nor*

When parts of a subject are joined by *or* or *nor,* the verb agrees with the nearer part.

Either the painter or the carpenter knows the cost.

The cabinets or the bookcases are too costly.

When one part of the subject is singular and the other is plural, the sentence will be awkward unless you put the plural part second.

Awkward Neither the owners nor the builder agrees.

Improved Neither the builder nor the owners agree.

11e ▪ *Everyone* and other indefinite pronouns

Indefinite pronouns° include *anybody, anyone, each, everybody, everyone, neither, no one, one,* and *somebody.* Most are singular in meaning and take singular verbs.

Something smells. Neither is right.

A few indefinite pronouns such as *all, any, none,* and *some* may take a singular or plural verb depending on whether the word they refer to is singular or plural.

All of the money is reserved for emergencies.

All of the funds are reserved for emergencies.

11f ▪ *Team* and other collective nouns

A collective noun° such as *team* or *family* takes a singular verb when the group acts as a unit.

The group agrees that action is necessary.

But when the group's members act separately, use a plural verb.

The old group have gone their separate ways.

11g ▪ *Who, which,* and *that*

When used as subjects, *who, which,* and *that* refer to another word in the sentence. The verb agrees with this other word.

Mayor Garber ought to listen to the people who work for her.

Bardini is the only aide who has her ear.

Bardini is one of the aides who work unpaid. [Of the aides who work unpaid, Bardini is one.]

Bardini is the only one of the aides who knows the community. [Of the aides, only one, Bardini, knows the community.]

11h ▪ *News* and other singular nouns ending in *-s*

Singular nouns° ending in *-s* include *athletics, economics, mathematics, measles, mumps, news, physics, politics,* and *statistics.*

After so long a wait, the news has to be good.

Statistics is required of psychology majors.

These words take plural verbs when they describe individual items rather than whole bodies of activity or knowledge.

The statistics prove him wrong.

11i ▪ Inverted word order

Is voting a right or a privilege?

Are a right and a privilege the same thing?

There are differences between them.

11j ▪ *Is, are,* and other linking verbs

Make a linking verb° agree with its subject, usually the first element in the sentence, not with other words referring to the subject.

The child's sole support is her court-appointed guardians.

Her court-appointed guardians are the child's sole support.

PRONOUNS

12 Pronoun Forms

A noun° or pronoun° changes form to show the reader how it functions in a sentence. These forms—called **cases**—are **subjective** (such as *I, she, they, man*), **objective** (such as *me, her, them, man*), and **possessive** (such as *my, her, their, man's*). A list of the case forms appears on pages 229–30.

Note Some grammar and style checkers can't spot any problems with pronoun form. Others do flag mistakes, but they also miss a lot. For instance, one checker spotted the error in *We asked whom would come* (should be *who*), but it overlooked *We dreaded them coming* (should be *their*). To catch possible errors, review your sentences on your own.

12a ▪ Compound subjects and objects: *she and I* vs. *her and me*

Subjects° and objects° consisting of two or more nouns and pronouns have the same case forms as they would if one pronoun stood alone.

Visit ***www.ablongman.com/littlebrown*** for added help and exercises on pronoun forms.

compound subject
She and Ming discussed the proposal.

compound object
The proposal disappointed her and him.

To test for the correct form, try one pronoun alone in the sentence. The case form that sounds correct is probably correct for all parts of the compound.

The prize went to [he, him] and [I, me].

The prize went to him.

The prize went to him and me.

12b ▪ Subject complements: *it was she*

Both a subject and a subject complement° appear in the same form—the subjective case.

subject complement
The one who cares most is she.

If this construction sounds stilted to you, use the more natural order: *She is the one who cares most.*

12c ▪ *Who* vs. *whom*

The choice between *who* and *whom* depends on the use of the word.

Questions

At the beginning of a question use *who* for a subject and *whom* for an object.

subject → Who wrote the policy? object ← Whom does it affect?

Test for the correct form by answering the question with the form of *he* or *she* that sounds correct. Then use the same form in the question.

[Who, Whom] does one ask?

One asks her.

Whom does one ask?

Subordinate clauses

In subordinate clauses° use *who* and *whoever* for all subjects, *whom* and *whomever* for all objects.

subject →
Give old clothes to whoever needs them.

object ←
I don't know whom the mayor appointed.

Test for the correct form by rewriting the subordinate clause as a sentence. Replace *who* or *whom* with the form of *he* or *she* that sounds correct. Then use the same form in the original subordinate clause.

Few people know [who, whom] they should ask.

They should ask her.

Few people know whom they should ask.

Note Don't let expressions such as *I think* and *she says* confuse you when they come between the subject *who* and its verb.

subject
He is the one who I think is best qualified.

12d ▪ Other constructions

We or *us* with a noun

The choice of *we* or *us* before a noun depends on the use of the noun.

object of preposition
Freezing weather is welcomed by us skaters.

subject
We skaters welcome freezing weather.

Pronoun in an appositive

An **appositive** is a word or word group that renames a noun or pronoun. Within an appositive the form of a pronoun depends on the function of the word the appositive renames.

object of verb
The class elected two representatives, DeShawn and me.

subject
Two representatives, DeShawn and I, were elected.

Pronoun after *than* or *as*

After *than* or *as* in a comparison, the form of a pronoun indicates what words may have been omitted. A subjective pronoun must be the subject of the omitted verb:

subject
Some critics like Glass more than she [does].

An objective pronoun must be the object of the omitted verb:

object
Some critics like Glass more than [they like] her.

Subject and object of an infinitive

An **infinitive** is the plain form° of the verb plus *to* (*to swim*). Both its object and its subject are in the objective form.

subject of infinitive

The school asked him to speak.

object of infinitive

Students chose to invite him.

Form before a gerund

A **gerund** is the *-ing* form of a verb used as a noun (*a runner's breathing*). Generally, use the possessive form of a pronoun or noun immediately before a gerund.

The coach disapproved of their lifting weights.

The coach's disapproving was a surprise.

13 Agreement of Pronoun and Antecedent

The word a pronoun refers to is its **antecedent.**

Homeowners fret over their tax bills.
antecedent — pronoun

For clarity, a pronoun should agree with its antecedent in person° (first, second, third), number° (singular, plural), and gender° (masculine, feminine, neuter).

Note Grammar and style checkers cannot help with agreement between pronoun and antecedent because they cannot recognize the intended relation between the two. You'll need to check for errors on your own.

13a ▪ Antecedents with *and*

The dean and my adviser have offered their help.

Note When *each* or *every* precedes the compound antecedent, the pronoun is singular.

Every girl and woman took her seat.

13b ▪ Antecedents with *or* or *nor*

When parts of an antecedent are joined by *or* or *nor*, the pronoun agrees with the nearer part.

Visit *www.ablongman.com/littlebrown* for added help and exercises on pronoun-antecedent agreement.

Tenants or owners must present their grievances.

Either the tenant or the owner will have her way.

When one subject is plural and the other singular, put the plural subject second to avoid awkwardness.

Neither the owner nor the tenants have made their case.

13c ▪ *Everyone, person,* and other indefinite words

Indefinite words do not refer to any specific person or thing. They include indefinite pronouns° (such as *anyone, everybody, everything, no one, somebody*) and generic nouns° (such as *person, individual, child, student*).

Most indefinite words are singular in meaning and take singular pronouns.

Everyone on the women's team now has her own locker.

Each of the men still has his own locker.

Though they are singular, indefinite words often seem to mean "many" or "all" rather than "one" and are mistakenly referred to with plural pronouns, as in *Everyone deserves their privacy.* Often, too, we mean indefinite words to include both masculine and feminine genders and thus resort to *they* instead of the **generic** ***he***—the masculine pronoun referring to both genders, which is generally regarded as sexist: *Everyone deserves his privacy*.

To achieve nonsexist agreement in such cases, you have several options:

- Change the indefinite word to a plural, and use a plural pronoun to match.

  Faulty Each athlete is entitled to his own locker.

  Revised All athletes are entitled to their own lockers. [*Is* changes to *are*, and *locker* changes to *lockers*.]

- Rewrite the sentence to omit the pronoun.

  Revised Each athlete is entitled to a locker.

- Use *he or she* (*him or her, his or her*) to refer to the indefinite word.

  Revised Each athlete is entitled to his or her own locker.

  He or she can be awkward, so avoid using it more than once in several sentences. Also avoid the combination *he/she,* which many readers do not accept.

13d ▪ *Team* and other collective nouns

Use a singular pronoun with *team, family, group,* or another collective noun° when referring to the group as a unit.

The committee voted to disband itself.

When referring to the individual members of the group, use a plural pronoun.

The old group have gone their separate ways.

14 Reference of Pronoun to Antecedent

If a pronoun° does not refer clearly to the word it substitutes for (its **antecedent**), readers will have difficulty grasping the pronoun's meaning.

Note Grammar and style checkers are not sophisticated enough to recognize unclear pronoun reference. For instance, a checker did not flag any of the confusing examples on this and the next page. You must proofread your work to spot unclear pronoun reference.

14a ▪ Single antecedent

When either of two nouns can be a pronoun's antecedent, the reference will not be clear.

Confusing Emily Dickinson is sometimes compared with Jane Austen, but she was quite different.

Revise such a sentence in one of two ways:

- Replace the pronoun with the appropriate noun.

  Clear Emily Dickinson is sometimes compared with Jane Austen, but Dickinson [or Austen] was quite different.

- Avoid repetition by rewriting the sentence. If you use the pronoun, make sure it has only one possible antecedent.

  Clear Despite occasional comparison, Emily Dickinson and Jane Austen were quite different.

Visit *www.ablongman.com/littlebrown* for added help and exercises on pronoun reference.

Clear — Though sometimes compared with her, Emily Dickinson was quite different from Jane Austen.

14b ▪ Close antecedent

A clause° beginning *who, which,* or *that* should generally fall immediately after the word it refers to.

Confusing — Jody found a dress in the attic that her aunt had worn.

Clear — In the attic Jody found a dress that her aunt had worn.

14c ▪ Specific antecedent

A pronoun should refer to a specific noun° or other pronoun.

Vague *this, that, which,* or *it*

This, that, which, or *it* should refer to a specific noun, not to a whole word group expressing an idea or situation.

Confusing — The British knew little of the American countryside, and they had no experience with the colonists' guerrilla tactics. This gave the colonists an advantage.

Clear — The British knew little of the American countryside, and they had no experience with the colonists' guerrilla tactics. This ignorance and inexperience gave the colonists an advantage.

Implied nouns

A pronoun cannot refer clearly to a noun that is merely implied by some other word or phrase, such as *news* in *newspaper* or *happiness* in *happy.*

Confusing — Cohen's report brought her a lawsuit.

Clear — Cohen was sued over her report.

Confusing — Her reports on psychological development generally go unnoticed outside it.

Clear — Her reports on psychological development generally go unnoticed outside the field.

Indefinite *it, they,* or *you*

It and *they* should have definite antecedents.

Confusing	In the average television drama they present a false picture of life.
Clear	The average television drama presents a false picture of life.

You should clearly mean "you, the reader," and the context must support such a meaning.

Inappropriate	In the fourteenth century you had to struggle simply to survive.
Revised	In the fourteenth century one [or a person or people] had to struggle to survive.

14d ▪ Consistency in pronouns

Within a sentence or a group of related sentences, pronouns should be consistent. You may shift pronouns unconsciously when you start with *one* and soon find it too stiff.

Inconsistent	One will find when reading that your concentration improves with practice, so that you comprehend more in less time.
Revised	You will find when reading that your concentration improves with practice, so that you comprehend more in less time.

Inconsistent pronouns also occur when singular shifts to plural: *Everyone who reads regularly will improve his or her* [not *their*] *speed.* See pages 49–50.

MODIFIERS

15 Adjectives and Adverbs

Adjectives modify nouns° (*good child*) and pronouns° (*special someone*). **Adverbs** modify verbs° (*see well*), adjec-

Visit *www.ablongman.com/littlebrown* for added help and exercises on adjectives and adverbs.

tives (*very happy*), other adverbs (*not very*), and whole word groups (*Otherwise, the room was empty*). The only way to tell if a modifier should be an adjective or an adverb is to determine its function in the sentence.

Note Grammar and style checkers will spot some but not all problems with misused adjectives and adverbs. For instance, a checker flagged *Some children suffer bad* and *Jenny did not feel nothing*. But it did not flag *Educating children good is everyone's focus*. You'll need to proofread your work on your own to be sure you've used adjectives and adverbs appropriately.

15a ▪ Adjective vs. adverb

Use only adverbs, not adjectives, to modify verbs, adverbs, or other adjectives.

Not They took each other serious. They related good.

But They took each other seriously. They related well.

15b ▪ Adjective with linking verb: *felt bad*

A modifier after a verb should be an adjective if it describes the subject,° an adverb if it describes the verb. In the first example below, the linking verb° *felt* connects the subject and an adjective describing the subject.

The sailors felt bad.
linking verb adjective

Some sailors fare badly in rough weather.
verb adverb

Good and *well* are frequently confused after verbs.

Decker trained well. [Adverb.]

She felt well. Her prospects were good. [Adjectives.]

15c ▪ Comparison of adjectives and adverbs

Comparison° allows adjectives and adverbs to show degrees of quality or amount by changing form: *red, redder, reddest; awful, more awful, most awful; quickly, less quickly, least quickly*. A dictionary will list the *-er* and *-est* endings if they can be used. Otherwise, use *more* and *most* or *less* and *least*.

Some modifiers are irregular, changing their spelling for comparison: for example, *good, better, best; many, more, most; badly, worse, worst*.

Comparisons of two or more than two

Use the *-er* form, *more,* or *less* when comparing two items. Use the *-est* form, *most,* or *least* when comparing three or more items.

> Of the two tests, the litmus is better.
> Of all six tests, the litmus is best.

Double comparisons

A double comparison combines the *-er* or *-est* ending with the word *more* or *most.* It is redundant.

> Chang was the wisest [not most wisest] person in town.
> He was smarter [not more smarter] than anyone else.

Complete comparisons

A comparison should be complete.

- The comparison should state a relation fully enough to ensure clarity.

Unclear	Car makers worry about their industry more than environmentalists.
Clear	Car makers worry about their industry more than environmentalists do.
Clear	Car makers worry about their industry more than they worry about environmentalists.

- The items being compared should in fact be comparable.

Illogical	The cost of an electric car is greater than a gasoline-powered car. [Illogically compares a cost and a car.]
Revised	The cost of an electric car is greater than the cost of [or that of] a gasoline-powered car.

15d ▪ Double negatives

In a **double negative** two negative words cancel each other out. Some double negatives are intentional, as *She was not unhappy* indicates with understatement that she was indeed happy. But most double negatives say the opposite of what is intended: *Jenny did not feel nothing* asserts that Jenny felt other than nothing, or something.

Faulty	The IRS cannot hardly audit all tax returns. None of its audits never touch many cheaters.
Revised	The IRS cannot audit all tax returns. Its audits never touch many cheaters.

15e ▪ Present and past participles as adjectives ESL

Both present participles° and past participles° may serve as adjectives: *a burning house, a burned house.* As in the examples, the two participles usually differ in the time they indicate.

But some present and past participles—those derived from verbs expressing feeling—can have altogether different meanings. The present participle refers to something that causes the feeling: *That was a frightening storm.* The past participle refers to something that experiences the feeling: *They quieted the frightened horses.* Similar pairs include the following:

annoying/annoyed
boring/bored
confusing/confused
exciting/excited
exhausting/exhausted
interesting/interested
pleasing/pleased
satisfying/satisfied
surprising/surprised
tiring/tired
troubling/troubled
worrying/worried

15f ▪ Articles: *a, an, the* ESL

Articles° usually trouble native English speakers only in the choice of *a* versus *an: a* for words beginning with consonant sounds (*a bridge*), *an* for words beginning with vowel sounds, including silent *h*'s (*an apple, an hour*).

For nonnative speakers, *a, an,* and *the* can be difficult, because many other languages use such words quite differently or not at all. In English, their uses depend on their context and the kinds of nouns they precede.

Singular count nouns

A **count noun** names something countable and can form a plural: *glass/glasses, mountain/mountains, child/children, woman/women.*

- *A* or *an* precedes a singular count noun when your reader does not already know its identity, usually because you have not mentioned it before.

  A scientist in our chemistry department developed a process to strengthen metals. [*Scientist* and *process* are being introduced for the first time.]

- *The* precedes a singular count noun that has a specific identity for your reader, usually because (1) you have mentioned it before, (2) you identify it immediately before or after you state it, (3) it is unique (the only one in existence), or (4) it refers to an institution or facility that is shared by the community.

A scientist in our chemistry department developed a process to strengthen metals. The scientist patented the process. [*Scientist* and *process* were identified before.]

The most productive laboratory is the research center in the chemistry department. [*Most productive* identifies *laboratory. In the chemistry department* identifies *research center.* And *chemistry department* is a shared facility.]

The sun rises in the east. [*Sun* and *east* are unique.]

Many men and women aspire to the presidency. [*Presidency* is a shared institution.]

Plural count nouns

A or *an* never precedes a plural noun. *The* does not precede a plural noun that names a general category. *The* does precede a plural noun that names specific representatives of a category.

Men and women are different. [*Men* and *women* name general categories.]

The women formed a team. [*Women* refers to specific people.]

Noncount nouns

A **noncount noun** names something that is not usually considered countable in English, and so it does not form a plural. Examples include the following:

Abstractions: confidence, democracy, education, equality, evidence, health, information, intelligence, knowledge, luxury, peace, pollution, research, success, supervision, truth, wealth, work

Food and drink: bread, flour, meat, milk, salt, water, wine

Emotions: anger, courage, happiness, hate, love, respect, satisfaction

Natural events and substances: air, blood, dirt, gasoline, gold, hair, heat, ice, oil, oxygen, rain, smoke, wood

Groups: clergy, clothing, equipment, furniture, garbage, jewelry, junk, legislation, mail, military, money, police

Fields of study: architecture, accounting, biology, business, chemistry, engineering, literature, psychology, science

A or *an* never precedes a noncount noun. *The* does precede a noncount noun that names specific representatives of a general category.

Vegetation suffers from drought. [*Vegetation* names a general category.]

The vegetation in the park withered or died. [*Vegetation* refers to specific plants.]

Note Many nouns are sometimes count nouns and sometimes noncount nouns.

> The library has a room for readers. [*Room* is a count noun meaning "walled area."]
>
> The library has room for reading. [*Room* is a noncount noun meaning "space."]

Proper nouns

A **proper noun** names a particular person, place, or thing and begins with a capital letter: *February, Joe Allen. A* or *an* never precedes a proper noun. *The* does occasionally, as with oceans (*the Pacific*), regions (*the Middle East*), rivers (*the Snake*), some countries (*the United States*), and some universities (*the University of Texas*).

> Garcia lives in Boulder, where he attends the University of Colorado.

16 Misplaced and Dangling Modifiers

For clarity, modifiers generally must fall close to the words they modify.

Note Grammar and style checkers do not recognize many problems with modifiers. For instance, a checker failed to flag the misplaced modifiers in *Gasoline high prices affect usually car sales* or the dangling modifier in *The vandalism was visible passing the building*. Proofread your work on your own to find and correct problems with modifiers.

16a ▪ Misplaced modifiers

A **misplaced modifier** falls in the wrong place in a sentence. It may be awkward, confusing, or even unintentionally funny.

Clear placement

Confusing	He served steak to the men on paper plates.
Revised	He served the men steak on paper plates.

Visit *www.ablongman.com/littlebrown* for added help and exercises on misplaced and dangling modifiers.

Confusing Many dogs are killed by automobiles and trucks roaming unleashed.

Revised Many dogs roaming unleashed are killed by automobiles and trucks.

Only and other limiting modifiers

Limiting modifiers include *almost, even, exactly, hardly, just, merely, nearly, only, scarcely,* and *simply.* They should fall immediately before the word or word group they modify.

Unclear They only saw each other during meals.

Revised They saw only each other during meals.

Revised They saw each other only during meals.

Infinitives and other grammatical units

Some grammatical units should generally not be split by long modifiers. For example, a long adverb° between subject° and verb° can be awkward and confusing.

Awkward Kuwait, after the first Gulf war ended in 1991, returned to normal.

Revised After the first Gulf war ended in 1991, Kuwait returned to normal.

A **split infinitive**—a modifier placed between *to* and the verb—can be especially awkward and will annoy many readers.

Awkward Farmers expected temperatures to not rise.

Revised Farmers expected temperatures not to rise.

A split infinitive may sometimes be unavoidable without rewriting, though it may still bother some readers.

Several US industries expect to more than triple their use of robots.

Order of adjectives ESL

English follows distinctive rules for arranging two or three adjectives before a noun. (A string of more than three adjectives before a noun is rare.) The adjectives follow the order shown on the next page.

Determiner	Opinion	Size or shape	Color	Origin	Material	Noun used as adjective	Noun
many						state	**laws**
	lovely		green	Thai			**birds**
a		square			wooden		**table**
all						business	**reports**
the			blue		litmus		**paper**

16b ▪ Dangling modifiers

A **dangling modifier** does not sensibly modify anything in its sentence.

Dangling Passing the building, the vandalism became visible.

Like most dangling modifiers, this one introduces a sentence, contains a verb form (*passing*), and implies but does not name a subject (whoever is passing). Readers assume that this implied subject is the same as the subject of the sentence (*vandalism*). When it is not, the modifier "dangles" unconnected to the rest of the sentence.

Revise dangling modifiers to achieve the emphasis you want.

- Rewrite the dangling modifier as a complete clause with its own stated subject and verb. Readers can accept different subjects when they are both stated.

Dangling Passing the building, the vandalism became visible.

Revised As we passed the building, the vandalism became visible.

- Change the subject of the sentence to a word the modifier properly describes.

Dangling Trying to understand the causes, vandalism has been extensively studied.

Revised Trying to understand the causes, researchers have extensively studied vandalism.

17 Sentence Fragments

A **sentence fragment** is part of a sentence that is set off as if it were a whole sentence by an initial capital letter and a final period or other end punctuation. Although writers occasionally use fragments deliberately and effectively, readers perceive most fragments as serious errors in standard English. Use the tests below to ensure that you have linked or separated your ideas both appropriately for your meaning and correctly, without creating sentence fragments.

Note Grammar and style checkers can spot many but not all sentence fragments. For instance, a checker flagged *The network growing* as a fragment but failed to flag *Uncountable numbers of sites on the Web.* Repair any fragments that your checker does find, but proofread your work yourself to ensure that it's fragment free.

ESL Some languages other than English allow the omission of the subject° or the verb.° Except in commands (*Close the door*), English always requires you to state the subject and verb.

17a ▪ Tests for fragments

A word group punctuated as a sentence should pass *all three* of the following tests. If it does not, it is a fragment and needs to be revised.

Test 1: Find the verb.

Some sentence fragments lack any verb form:°

Fragment	Uncountable numbers of sites on the Web.
Revised	Uncountable numbers of sites make up the Web.

The verb in a complete sentence must be able to change form as on the left of the following chart. A verb form that cannot change this way (as on the right) cannot serve as a sentence verb.

Visit *www.ablongman.com/littlebrown* for added help and exercises on sentence fragments.

	Complete sentences	Sentence fragments
Singular	The network grows.	The network growing.
Plural	Networks grow.	Networks growing.
Present	The network grows.	The network growing.
Past	The network grew.	
Future	The network will grow.	

(See also p. 34 on the use of helping verbs° to prevent sentence fragments.)

Test 2: Find the subject.

The subject of the sentence will usually come before the verb. If there is no subject, the word group is probably a fragment.

Fragment And has enormous popular appeal.

Revised And the Web has enormous popular appeal.

Note Commands, in which the subject *you* is understood, are not sentence fragments: [*You*] *Experiment with the Web.*

Test 3: Make sure the clause is not subordinate.

A **subordinate clause** begins with either a subordinating conjunction° (such as *because, if, when*) or a relative pronoun° (*who, which, that*). Subordinate clauses serve as parts of sentences, not as whole sentences.

Fragment The Internet was greatly improved by Web technology. Which allows users to move easily between sites.

Revised The Internet was greatly improved by Web technology, which allows users to move easily between sites. [The subordinate clause joins a main clause in a complete sentence.]

Revised The Internet was greatly improved by Web technology. It allows users to move easily between sites. [Substituting *It* for *Which* makes the subordinate clause into a complete sentence.]

Note Questions beginning *who, whom,* or *which* are not sentence fragments: *Who rattled the cage?*

17b ▪ Revision of fragments

Correct sentence fragments in one of the two ways shown in the revised examples above, depending on the importance of the information in the fragment:

- Combine the fragment with the appropriate main clause. The information in the fragment will then be subordinated to that in the main clause.

Fragment — The Web is easy to use. Loaded with links and graphics.

Revised — The Web, loaded with links and graphics, is easy to use.

- Rewrite the fragment as a complete sentence. The information in the fragment will then have the same importance as that in other complete sentences.

Fragment — The Internet was a true innovation. Because no expansive computer network existed before it.

Revised — The Internet was a true innovation. No expansive computer network existed before it. [Deleting *Because* makes the fragment into a complete sentence.]

18 Comma Splices and Fused Sentences

When you combine two complete sentences (main clauses°) in one sentence, you need to give readers a clear signal that one clause is ending and the other beginning. In a **comma splice** two main clauses are joined (or spliced) only by a comma, which is usually too weak to signal the link between main clauses.

Comma splice — The ship was huge, its mast stood eighty feet high.

In a **fused sentence** (or **run-on sentence**) the clauses are not separated at all.

Fused sentence — The ship was huge its mast stood eighty feet high.

Note Grammar and style checkers can detect many comma splices, but they will miss most fused sentences. For example, a checker flagged *Money is tight, we need to spend carefully* but not *Money is tight we need to spend carefully.* A checker may also question sentences that are actually correct, such as *Money being tighter now than before, we need to spend more carefully.* Verify that revision is actually needed on any flagged sentence, and read your work carefully on your own to be sure it is correct.

Visit *www.ablongman.com/littlebrown* for added help and exercises on comma splices and fused sentences.

18a ▪ Main clauses without *and, but, or, nor, for, so, yet*

Two main clauses in a sentence are usually separated with a comma and a coordinating conjunction° such as *and* or *but.* These signals tell readers to expect another main clause. When one or both signals are missing, the sentence may be confusing and may require rereading. Revise it in one of the following ways:

- Insert a coordinating conjunction when the ideas in the main clauses are closely related and equally important.

Comma splice — Some laboratory-grown foods taste good, they are nutritious.

Revised — Some laboratory-grown foods taste good, and they are nutritious.

In a fused sentence insert a comma and a coordinating conjunction.

Fused sentence — Chemists have made much progress they still have a way to go.

Revised — Chemists have made much progress, but they still have a way to go.

- Insert a semicolon between clauses if the relation between the ideas is very close and obvious without a conjunction.

Comma splice — Good taste is rare in laboratory-grown vegetables, they are usually bland.

Revised — Good taste is rare in laboratory-grown vegetables; they are usually bland.

- Make the clauses into separate sentences when the ideas expressed are only loosely related.

Comma splice — Chemistry has contributed to our understanding of foods, many foods such as wheat and beans can be produced in the laboratory.

Revised — Chemistry has contributed to our understanding of foods. Many foods such as wheat and beans can be produced in the laboratory.

- Subordinate one clause to the other when one idea is less important than the other. The subordinate clause will modify something in the main clause.

Comma splice — The vitamins are adequate, the flavor and color are deficient.

Revised Even though the vitamins are adequate, the flavor and color are deficient.

18b ▪ Main clauses related by *however, for example,* and so on

Two kinds of words can describe how one main clause relates to another: conjunctive adverbs,° such as *however, instead, meanwhile,* and *thus*; and other transitional expressions,° such as *even so, for example, in fact,* and *of course.* Two main clauses related by all conjunctive adverbs and most transitional expressions must be separated by a period or by a semicolon. The connecting word or phrase is also generally set off by a comma or commas.

Comma splice Most Americans refuse to give up unhealthful habits, consequently our medical costs are higher than those of many other countries.

Revised Most Americans refuse to give up unhealthful habits. Consequently, our medical costs are higher than those of many other countries.

Revised Most Americans refuse to give up unhealthful habits; consequently, our medical costs are higher than those of many other countries.

To test whether a word or phrase is a conjunctive adverb or transitional expression, try moving it around in its clause. Either kind of word can move, whereas other kinds (such as *and* or *because*) cannot.

Most Americans refuse to give up unhealthful habits; our medical costs, consequently, are higher than those of many other countries.

PART III

Punctuation

CHECKLIST

Punctuation

This checklist focuses on the most troublesome punctuation marks and uses, showing correctly punctuated sentences with brief explanations. For the other marks and other uses covered in this part, see the contents inside the back cover.

Comma

☑ Subways are convenient, but they are costly to build.
Subways are convenient but costly.
[With *and, but,* etc., only between main clauses. See opposite.]

☑ Because of their cost, new subways are rarely built.
[With an introductory element. See opposite.]

☑ Light rail, which is less costly, is often more feasible.
Those who favor mass transit often propose light rail.
[With a nonessential element, not with an essential element. See p. 70.]

☑ In a few older cities, commuters can choose from subways, buses, light rail, and railroads.
[Separating items in a series. See p. 72.]

Semicolon

☑ She chose carpentry; she wanted manual work.
She had a law degree; however, she became a carpenter.
[Between main clauses not joined by *and, but,* etc., and those joined by *however, for example,* etc. See p. 74.]

Colon

☑ The school has one goal: to train businesspeople.
[With a main clause to introduce information. See p. 76.]

Apostrophe

☑ Bill Smith's dog saved the life of the Smiths' grandchild.
[Showing possession: with *-'s* for singular nouns; with *-'* alone for plural nouns ending in *-s*. See pp. 77–78.]

☑ Its [for The dog's] bark warned the family.
It's [for It is] an intelligent dog.
[Not with possessive pronouns, only with contractions. See p. 78]

19 The Comma

The comma helps to separate sentence elements and to prevent misreading. Its main uses (and misuses) appear below.

Note Grammar and style checkers will ignore many comma errors. For example, a checker failed to catch the missing commas in *We cooked lasagna spinach and apple pie* and the misused comma in *The trip was short but, the weather was perfect.* Revise any errors that your checker does catch, but you'll have to proofread your work on your own to find and correct most errors.

19a ▪ Comma with *and, but, or, nor, for, so, yet*

Between main clauses

Use a comma before *and, but, or, nor, for, so,* and *yet* (the coordinating conjunctions°) when they link main clauses.°

> Banks offer many services, but they could do more.
>
> Many banks offer investment advice, and they may help small businesses establish credit.

Note The comma goes before, not after, the coordinating conjunction.

Not between words, phrases, or subordinate clauses

Generally, do not use a comma before *and, but, or,* and *nor* when they link pairs of words, phrases,° or subordinate clauses°—that is, elements other than main clauses.

Not One bank established special accounts for older depositors, and counseled them on investments.

But One bank established special accounts for older depositors and counseled them on investments.

19b ▪ Comma with introductory elements

Use a comma after most elements that begin sentences and are distinct from the main clause.

Visit *www.ablongman.com/littlebrown* for added help and exercises on the comma.

When a new century nears, futurists multiply.

Fortunately, some news is good.

You may omit the comma after a short introductory element if there's no risk that the reader will run the introductory element and main clause together: *By 2010 we may have reduced pollution.*

Note The subject° of a sentence is not an introductory element but a part of the main clause. Thus, do not use a comma to separate the subject and its verb.

Not Some pessimists, may be disappointed.

But Some pessimists may be disappointed.

19c ▪ Comma or commas with interrupting and concluding elements

Use a comma or commas to set off information that could be deleted without altering the basic meaning of the sentence.

Note When such optional information falls in the middle of the sentence, be sure to use one comma *before* and one *after* it.

Around nonessential elements

A **nonessential** (or **nonrestrictive**) **element** adds information about a word in the sentence but does not limit (or restrict) the word to a particular individual or group. Omitting the underlined element from any sentence below would remove incidental details but would not affect the sentence's basic meaning.

Nonessential modifiers

Hai Nguyen, who emigrated from Vietnam, lives in Denver.

His company, which is ten years old, studies air and water pollution.

Nguyen's family lives in Baton Rouge and Chicago, even though he lives in Denver.

Nonessential appositives

Appositives are words or word groups that rename nouns.

Hai Nguyen's work, advanced research into air pollution, keeps him in Denver.

His wife, Tina Nguyen, reports for a newspaper in Chicago.

Not around essential elements

Do not use commas to set off **essential** (or **restrictive**) **elements,** modifiers and appositives that contain information essential to the meaning of the sentence. Omitting the underlined element from any sentence below would alter the meaning substantially, leaving the sentence unclear or too general.

Essential modifiers

People who join recycling programs rarely complain about the extra work.

The programs that succeed are often staffed by volunteers.

Most people recycle because they believe they have a responsibility to the earth.

Essential appositives

The label "Recycle" on products becomes a command.
The activist Susan Bower urges recycling.
The book *Efficient Recycling* provides helpful tips.

Around absolute phrases

An **absolute phrase** consists usually of the *-ing* form of a verb plus a subject for the verb. The phrase modifies the whole main clause of the sentence.

Health insurance, its cost always rising, is a concern for many students.

Around transitional or parenthetical expressions

A transitional expression° such as *however, for example,* and *of course* forms a link between ideas. It is nonessential and is usually set off with a comma or commas.

Most students at the city colleges, for example, have no health insurance.

A parenthetical expression° provides supplementary information not essential for meaning—for instance, *fortunately, to be frank,* and *published in 1990.* It can be enclosed in parentheses (see p. 84) or, with more emphasis, in commas.

Some schools, it seems, do not offer group insurance.

Note Do not add a comma after a coordinating conjunction° (*and, but,* and so on) or a subordinating conjunction° (*although, because,* and so on). To distinguish between these words and transitional or parenthetical expressions, try moving the word or expression around in its clause. Transitional or parenthetical expressions can be moved; coordinating and subordinating conjunctions cannot.

Around phrases of contrast

Students may focus on the cost of health care, not their health.

Around *yes* and *no*

All schools should agree that, yes, they will provide at least minimal insurance at low cost.

Around words of direct address

Heed this lesson, readers.

19d ▪ Commas with series

Between series items

Use commas to separate the items in lists or series.

The names Belial, Beelzebub, and Lucifer sound ominous.

The comma before the last item in a series (before *and*) is optional, but it is never wrong and it is usually clearer.

Not around series

Do not use a comma *before* or *after* a series.

Not The skills of, agriculture, herding, and hunting, sustained the Native Americans.

But The skills of agriculture, herding, and hunting sustained the Native Americans.

19e ▪ Comma with adjectives

Between equal adjectives

Use a comma between two or more adjectives° when each one modifies the same word equally. As a test, such adjectives could be joined by *and*.

The book had a worn, cracked binding.

Not between unequal adjectives

Do not use a comma between adjectives when one forms a unit with the modified word. As a test, the two adjectives could not sensibly be joined by *and*.

The house overflowed with ornate electric fixtures.
Among the junk in the attic was one lovely vase.

19f ▪ Commas with dates, addresses, place names, numbers

When they appear within sentences, elements punctuated with commas are also ended with commas.

Dates

July 4, 1776, was the day the Declaration was signed. [Note that commas appear before *and* after the year.]

The United States entered World War II in December 1941. [No comma is needed between a month or season and a year.]

Addresses and place names

Use the address 806 Ogden Avenue, Swarthmore, Pennsylvania 19081, for all correspondence. [No comma is needed between the state name and zip code.]

Numbers

The new assembly plant cost $7,525,000.
A kilometer is 3,281 feet [*or* 3281 feet].

19g ▪ Commas with quotations

A comma or commas usually separate a quotation from a signal phrase that identifies the source, such as *she said* or *he replied.*

Eleanor Roosevelt said, "You must do the thing you think you cannot do."

"Knowledge is power," wrote Francis Bacon.

"You don't need a weatherman," sings Bob Dylan, "to know which way the wind blows."

Do not use a comma when the signal phrase interrupts the quotation between main clauses.° Instead, follow the signal phrase with a semicolon or period.

"That part of my life was over," she wrote; "his words had sealed it shut."

"That part of my life was over," she wrote. "His words had sealed it shut."

20 The Semicolon

The semicolon separates equal and balanced sentence elements, usually main clauses.°

Note A grammar and style checker can spot a few errors in the use of semicolons. For example, a checker suggested using a semicolon after *perfect* in *The set was perfect, the director had planned every detail,* thus correcting a comma splice.° But it missed the incorrect semicolon in *The set was perfect; deserted streets, dark houses, and gloomy mist* (a colon would be correct). To find semicolon errors, you'll need to proofread on your own.

20a ▪ Semicolon between main clauses not joined by *and, but, or, nor,* etc.

20b

Between main clauses

Use a semicolon between main clauses° that are not connected by *and, but, or, nor, for, so,* or *yet* (the coordinating conjunctions°).

> Increased taxes are only one way to pay for programs**;** cost cutting also frees up money.

Not between main clauses and subordinate elements

Do not use a semicolon between a main clause and a subordinate element, such as a subordinate clause° or a phrase.°

> Not According to African authorities; Pygmies today number only about 35,000.
>
> But According to African authorities**,** Pygmies today number only about 35,000.

> Not Anthropologists have campaigned; for the protection of the Pygmies' habitat.
>
> But Anthropologists have campaigned for the protection of the Pygmies' habitat.

20b ▪ Semicolon with *however, for example,* and so on

Use a semicolon between main clauses° that are related by two kinds of words: conjunctive adverbs,° such as

Visit *www.ablongman.com/littlebrown* for added help and an exercise on the semicolon.

however, indeed, therefore, and *thus;* and other transitional expressions,° such as *after all, for example, in fact,* and *of course.*

> Blue jeans have become fashionable all over the world**;** however, the American originators still wear more jeans than anyone else.

A conjunctive adverb or transitional expression may move around within its clause, so the semicolon will not always come just before the adverb or expression. The adverb or expression itself is usually set off with a comma or commas.

> Blue jeans have become fashionable all over the world**;** the American originators**,** however**,** still wear more jeans than anyone else.

20c ▪ Semicolons with series

Between series items

Use semicolons (rather than commas) to separate items in a series when the items contain commas.

> The custody case involved Amy Dalton, the child**;** Ellen and Mark Dalton, the parents**;** and Ruth and Hal Blum, the grandparents.

Not before a series

Do not use a semicolon to introduce a series. (Use a colon or a dash instead.)

> Not Teachers have heard all sorts of reasons why students do poorly; psychological problems, family illness, too much work, too little time.
>
> But Teachers have heard all sorts of reasons why students do poorly**:** psychological problems, family illness, too much work, too little time.

21 The Colon

The colon is mainly a mark of introduction, but it has a few other conventional uses as well.

Visit *www.ablongman.com/littlebrown* for added help and an exercise on the colon.

Note Most grammar and style checkers cannot recognize missing or misused colons. For example, a checker failed to flag *The goal is: to raise $1 million* (no colon is needed) or *The President promised the following, lower taxes, stronger schools, and greater prosperity* (a colon, not a comma, should come after *following*). You'll have to check for colon errors yourself.

21a ▪ Colon for introduction

At end of main clause

The colon ends a main clause° and introduces various additions:

Soul food has a deceptively simple definition**:** the ethnic cooking of African Americans. [Introduces an explanation.]

At least three soul food dishes are familiar to most Americans**:** fried chicken, barbecued spareribs, and sweet potatoes. [Introduces a series.]

Soul food has one disadvantage**:** fat. [Introduces an appositive.°]

One soul food chef has a solution**:** "Instead of using ham hocks to flavor beans, I use smoked turkey wings. The soulful, smoky taste remains, but without all the fat of pork." [Introduces a long quotation.]

Not inside main clause

Do not use a colon inside a main clause, especially after *such as* or a verb.

Not The best-known soul food dish is: fried chicken. Many Americans have not tasted delicacies such as: chitlins and black-eyed peas.

But The best-known soul food dish is fried chicken. Many Americans have not tasted delicacies such as chitlins and black-eyed peas.

21b ▪ Colon with salutations of business letters, titles and subtitles, and divisions of time

Salutation of a business letter

Dear Ms. Burak**:**

Title and subtitle

*Anna Freud***:** *Her Life and Work*

Time

12**:**26 6**:**00

22 The Apostrophe

The apostrophe (') appears as part of a word to indicate possession, the omission of one or more letters, or (in a few cases) plural number.

Note Grammar and style checkers have mixed results in recognizing apostrophe errors: they may catch missing apostrophes in contractions (*isnt*) but overlook *boys* for *boy's* or *it's* for *its*. The checkers can identify some apostrophe errors in possessives but will overlook others and may flag correct plurals. Instead of relying on your checker, try using your word processor's Search or Find function to hunt for all words you have ended in *-s*. Then check them to ensure that they correctly omit or include apostrophes and that needed apostrophes are in the right places.

22a ▪ Apostrophe with possessives

The **possessive** form of a word indicates that it owns or is the source of another word: *the dog's hair, everyone's hope*. For nouns° and indefinite pronouns,° such as *everyone*, the possessive form always includes an apostrophe and often an *-s*.

Note The apostrophe or apostrophe-plus-*s* is an *addition*. Before this addition, always spell the name of the owner or owners without dropping or adding letters.

Singular words: Add *-'s*.

Bill Boughton's skillful card tricks amaze children.

Anyone's eyes would widen.

The *-'s* ending for singular words usually pertains to singular words ending in *-s*.

Sandra Cisneros's work is highly regarded.

The business's customers filed suit.

However, some writers add only the apostrophe to singular words ending in *-s*, especially when the additional *s* would make the word difficult to pronounce (*Moses'*) or when the name sounds like a plural (*Rivers'*).

Visit *www.ablongman.com/littlebrown* for added help and exercises on the apostrophe.

Plural words ending in *-s*: Add *-'* only.

Workers' incomes have fallen slightly over the past year.

Many students take several years' leave after high school.

The Murphys' son lives at home.

Plural words not ending in *-s*: Add *-'s*.

Children's educations are at stake.

We need to attract the media's attention.

Compound words: Add *-'s* only to the last word.

The brother-in-law's business failed.

Taxes are always somebody else's fault.

Two or more owners: Add *-'s* depending on possession.

Zimbale's and Mason's comedy techniques are similar. [Each comedian has his own technique.]

The child recovered despite her mother and father's neglect. [The mother and father were jointly neglectful.]

22c

22b ▪ Misuses of the apostrophe

Not with plural nouns°

Not The unleashed dog's belonged to the Jones'.

But The unleashed dogs belonged to the Joneses.

Not with singular verbs°

Not The subway break's down less often now.

But The subway breaks down less often now.

Not with possessives of personal pronouns°

Not The car is her's, not their's. It's color is red.

But The car is hers, not theirs. Its color is red.

Note Don't confuse possessive pronouns and contractions: *its, your, their,* and *whose* are possessives. *It's, you're, they're,* and *who's* are contractions. See below.

22c ▪ Apostrophe with contractions

A **contraction** replaces one or more letters, numbers, or words with an apostrophe.

it is	it's	cannot	can't
you are	you're	does not	doesn't
they are	they're	were not	weren't
who is	who's	class of 1997	class of '97

Note The contractions *it's, you're, they're,* and *who's* are easily confused with the possessive pronouns° *its, your, their,* and *whose.* To avoid misusing any of these words, search for all of them in your drafts (a word processor can help with this search). Then test for correctness:

- Do you intend the word to contain the sentence verb *is* or *are,* as in *It is a shame, They are to blame, You are right, Who is coming?* Then use an apostrophe: *it's, they're, you're, who's.*
- Do you intend the word to indicate possession, as in *Its tail was wagging, Their car broke down, Your eyes are blue, Whose book is that?* Then don't use an apostrophe.

22d ▪ Apostrophe with plural abbreviations, dates, and words or characters named as words

You'll sometimes see apostrophes used to form the plurals of abbreviations (*BA's*), dates (*1900's*), and words or characters used as words (*but's*). However, current style guides recommend against the apostrophe in these cases.

BAs PhDs
1990s 2000s

The sentence has too many *buts* [or buts].
Two *3*s [or 3s] end the zip code.

Note Italicize or underline a word or character named as a word (see p. 98), but not the added *-s*.

23 Quotation Marks

Quotation marks—either double (" ") or single (' ')—mainly enclose direct quotations from speech and from writing.

This chapter treats the main uses of quotation marks. Additional issues with quotations are discussed elsewhere in this book:

- Punctuating *she said* and other signal phrases with quotations (p. 73).

Visit *www.ablongman.com/littlebrown* for added help and an exercise on quotation marks.

- Altering quotations using the ellipsis mark (pp. 85–87) or brackets (p. 87).
- Quoting sources versus paraphrasing or summarizing them (pp. 128–30).
- Integrating quotations into your text (pp. 130–35).
- Avoiding plagiarism when quoting (pp. 136–40).
- Formatting long prose quotations and poetry quotations in MLA style (p. 160), APA style (p. 176), or Chicago style (p. 189).

Note Always use quotation marks in pairs, one at the beginning of a quotation and one at the end. Some grammar and style checkers will help you use quotation marks in pairs by flagging a lone mark. Others will identify where other punctuation falls incorrectly inside or outside quotation marks. For example, a checker flagged the comma outside the closing quotation mark in *"I must go", she said* (the comma should fall inside the quotation mark). But you'll need to proofread carefully yourself to catch the many possible errors in punctuating quotations that checkers cannot recognize.

23a ▪ Quotation marks with direct quotations

Double quotation marks

A **direct quotation** reports what someone said or wrote, in the exact words of the original.

> "Life," said the psychoanalyst Karen Horney, "remains a very efficient therapist."

Note Do not use quotation marks with an **indirect quotation,** which reports what someone said or wrote but not in the exact words of the original: *Karen Horney remarked that life is a good therapist.*

Single quotation marks

Use single quotation marks to enclose a quotation within a quotation.

> "In formulating any philosophy," Woody Allen writes, "the first consideration must always be: What can we know? Descartes hinted at the problem when he wrote, 'My mind can never know my body, although it has become quite friendly with my leg.'"

Dialog

When quoting a conversation, put the speeches in double quotation marks and begin a new paragraph for each speaker.

> "What shall I call you? Your name?" Andrews whispered rapidly, as with a high squeak the latch of the door rose.
>
> "Elizabeth," she said. "Elizabeth."
>
> —Graham Greene, *The Man Within*

23b ▪ Quotation marks with titles of works

Do not use quotation marks for the titles of your own papers. Within your text, however, use quotation marks to enclose the titles of works that are published or released within larger works (see below). Use underlining or italics for all other titles (see p. 97).

Short story
"The Gift of the Magi"

Short poem
"Her Kind"

Article in a periodical
"Does 'Scaring' Work?"

Page or document on a Web site
"Reader's Page" (on the site Friends of Prufrock)

Essay
"Joey: A 'Mechanical Boy'"

Song
"Satisfaction"

Episode of a television or radio program
"The Mexican Connection" (on 60 Minutes)

Subdivision of a book
"The Mast Head" (Chapter 35 of Moby-Dick)

Note APA and CSE styles do not use quotation marks for titles within source citations. See pages 166 and 193.

23c ▪ Quotation marks with words used in a special sense

On movie sets movable "wild walls" make a one-walled room seem four-walled on film.

Avoid using quotation marks to excuse slang or to express irony—that is, to indicate that you are using a word with a different or even opposite meaning than usual.

Not Americans "justified" their treatment of the Indians.

But Americans attempted to justify their treatment of the Indians.

Note Use underlining or italics to highlight words you are defining. (See p. 98.)

23d ▪ Quotation marks with other punctuation

Commas and periods: Inside quotation marks

Jonathan Swift wrote a famous satire, "A Modest Proposal," in 1729.

"Swift's 'A Modest Proposal,'" wrote one critic, "is so outrageous that it cannot be believed."

Exception When a parenthetical source citation immediately follows a quotation, place any comma or period *after* the citation:

One critic calls the essay "outrageous" (Olms 26).

Colons and semicolons: Outside quotation marks

A few years ago the slogan in elementary education was "learning by playing"; now educators stress basic skills.

We all know the meaning of "basic skills": reading, writing, and arithmetic.

Dashes, question marks, and exclamation points: Inside quotation marks only if part of the quotation

When a dash, exclamation point, or question mark is part of the quotation, place it *inside* quotation marks. Don't use any other punctuation, such as a period or comma.

"But must you—" Marcia hesitated, afraid of the answer.

"Go away!" I yelled.

Did you say, "Who is she?" [When both your sentence and the quotation would end in a question mark or exclamation point, use only the mark in the quotation.]

When a dash, question mark, or exclamation point applies only to the larger sentence, not to the quotation, place it *outside* quotation marks—again, with no other punctuation.

Betty Friedan's question in 1963—"Who knows what women can be?"—encouraged others to seek answers.

Who said, "Now cracks a noble heart"?

24 End Punctuation

End a sentence with one of three punctuation marks: a period, a question mark, or an exclamation point.

Note Do not rely on a grammar and style checker to identify missing or misused end punctuation. Although a

Visit *www.ablongman.com/littlebrown* for added help and an exercise on end punctuation.

checker may flag missing question marks after direct questions or incorrect combinations of marks, it cannot do much else.

24a ▪ Period for most sentences and some abbreviations

Statement	Mild command
The airline went bankrupt**.**	Think of the possibilities**.**

Indirect question°

The article asks how we can improve math education**.**
It asks what cost we are willing to pay**.**

Abbreviations

Use periods with abbreviations that end in small letters. Otherwise, omit periods from abbreviations.

Dr.	Mr., Mrs.	e.g.	Feb.	ft.
St.	Ms.	i.e.	p.	a.m., p.m.
PhD	BC, AD	USA	IBM	JFK
BA	AM, PM	US	USMC	AIDS

Note When a sentence ends in an abbreviation with a period, don't add a second period: *My first class is at 8 a.m.*

24b ▪ Question mark for direct questions°

What is the result**?**
What is the difference between those proposals**?**

24c ▪ Exclamation point for strong statements and commands

No**!** We must not lose this election**!**
Stop the car**!**

Note Use exclamation points sparingly to avoid seeming overly dramatic.

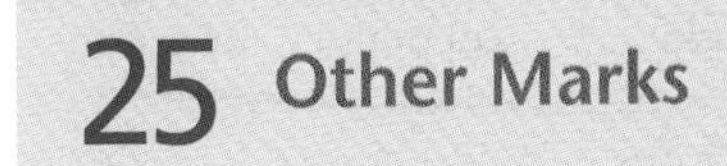

25 Other Marks

The other marks of punctuation are the dash, parentheses, the ellipsis mark, brackets, and the slash.

Visit *www.ablongman.com/littlebrown* for added help and exercises on the punctuation marks covered in this chapter.

Note Many grammar and style checkers will flag a lone parenthesis or bracket so that you can match it with another parenthesis or bracket. But most checkers cannot recognize other misuses of the marks covered here. You'll need to proofread your papers carefully for errors.

25a ▪ Dash or dashes: Shifts and interruptions

The dash (—) punctuates sentences, while the hyphen (-) punctuates words. Form a dash with two hyphens (--)—no extra space before, after, or between the hyphens. Or use the character called an em-dash on your word processor.

Note Be sure to use a pair of dashes when a shift or interruption falls in the middle of a sentence.

Shift in tone or thought

The novel—if one can call it that—appeared in 1994.

25b

Nonessential element

The qualities Monet painted—sunlight, shadows, deep colors—abounded near the rivers and gardens he used as subjects. [Commas may also set off nonessential elements. See p. 70.]

Introductory series

Shortness of breath, skin discoloration, persistent indigestion—all these may signify cancer.

Concluding series or explanation

The patient undergoes a battery of tests—CAT scan, ultrasound, and biopsy. [A colon may also set off a concluding series. See p. 76.]

25b ▪ Parentheses: Nonessential elements

Parentheses always come in pairs, one before and one after the punctuated material.

Parenthetical expressions

Parentheses de-emphasize explanatory or supplemental words or phrases. (Commas emphasize these expressions more and dashes still more.)

The population of Philadelphia (now about 1.5 million) has declined since 1950.

Don't put a comma before an opening parenthesis. After a parenthetical expression, place any punctuation *outside* the closing parenthesis.

Not Philadelphia's population compares with Houston's, (just over 1.6 million.)

But Philadelphia's population compares with Houston's (just over 1.6 million).

If a complete sentence falls within parentheses, place the period *inside* the closing parenthesis.

The population of Philadelphia has been dropping. (See p. 77 for population data since 1950.)

Labels for lists within text

Outside the Middle East, the countries with the largest oil reserves are (1) Venezuela (63 billion barrels), (2) Russia (57 billion barrels), and (3) Mexico (51 billion barrels).

Do not use parentheses for such labels when you set a list off from your text.

25c ▪ Ellipsis mark: Omissions from quotations

The ellipsis mark, consisting of three spaced periods (. . .), generally indicates an omission from a quotation. The following examples quote from or refer to this passage about environmentalism:

Original quotation

"At the heart of the environmentalist world view is the conviction that human physical and spiritual health depends on sustaining the planet in a relatively unaltered state. Earth is our home in the full, genetic sense, where humanity and its ancestors existed for all the millions of years of their evolution. Natural ecosystems—forests, coral reefs, marine blue waters—maintain the world exactly as we would wish it to be maintained. When we debase the global environment and extinguish the variety of life, we are dismantling a support system that is too complex to understand, let alone replace, in the foreseeable future."
—Edward O. Wilson,
"Is Humanity Suicidal?"

1. Omission of the middle of a sentence

Wilson observes, "Natural ecosystems . . . maintain the world exactly as we would wish it to be maintained."

2. Omission of the end of a sentence, without source citation

Wilson writes, "Earth is our home. . . ." [The sentence period, closed up to the last word, precedes the ellipsis mark.]

3. Omission of the end of a sentence, with source citation

Wilson writes, "Earth is our home . . ." (27). [The sentence period follows the source citation.]

4. Omission of parts of two or more sentences

Wilson writes, "At the heart of the environmentalist world view is the conviction that human physical and spiritual health depends on sustaining the planet . . . where humanity and its ancestors existed for all the millions of years of their evolution."

5. Omission of one or more sentences

As Wilson puts it, "At the heart of the environmentalist world view is the conviction that human physical and spiritual health depends on sustaining the planet in a relatively unaltered state. . . . When we debase the global environment and extinguish the variety of life, we are dismantling a support system that is too complex to understand, let alone replace, in the foreseeable future."

6. Omission from the middle of a sentence through the end of another sentence

"Earth is our home. . . . When we debase the global environment and extinguish the variety of life, we are dismantling a support system that is too complex to understand, let alone replace, in the foreseeable future."

7. Omission of the beginning of a sentence, leaving a complete sentence

a. Bracketed capital letter

"[H]uman physical and spiritual health," Wilson writes, "depends on sustaining the planet in a relatively unaltered state." [No ellipsis mark is needed because the brackets around the *H* indicate that the letter was not capitalized originally and thus that the beginning of the sentence has been omitted.]

b. Small letter

According to Wilson, "human physical and spiritual health depends on sustaining the planet in a relatively unaltered state." [No ellipsis mark is needed because the small *h* indicates that the beginning of the sentence has been omitted.]

c. Capital letter from the original

Hami comments, ". . . Wilson argues eloquently for the environmentalist world view." [An ellipsis mark *is* needed because the quoted part of the sentence begins with a capital letter and it is not clear that the beginning of the original sentence has been omitted.]

8. Use of a word or phrase

Wilson describes the earth as "our home." [No ellipsis mark needed.]

Note these features of the examples:

- Use an ellipsis mark when it is not otherwise clear that you have left out material from the source, as when you omit one or more sentences (examples 5 and 6) or when the words you quote form a complete sentence that is different in the original (examples 1–4 and 7c).
- You don't need an ellipsis mark when it is obvious that you have omitted something, such as when capitalization indicates omission (examples 7a and 7b) or when a phrase clearly comes from a larger sentence (example 8).
- After a grammatically complete sentence, an ellipsis mark follows the sentence period (examples 2, 5, and 6) *except* when a parenthetical source citation follows the quotation (example 3). Then the sentence period falls after the citation.

If your quotation omits one or more lines of poetry or paragraphs of prose, show the ommission with an entire line of ellipsis marks across the full width of the quotation.

25d ▪ Brackets: Changes in quotations

Though they have specialized uses in mathematical equations, brackets mainly indicate that you have altered a quotation to fit it into your sentence.

"[T]hat Texaco station [just outside Chicago] is one of the busiest in the nation," said a company spokesperson.

25e ▪ Slash: Options, breaks in poetry lines, and URLs

Option

Some teachers oppose pass/fail courses.

Break in poetry lines run into your text

Many readers have sensed a reluctant turn away from death in Frost's lines "The woods are lovely, dark and deep, / But I have promises to keep."

URL

http://www.stanford.edu/depts/spc/spc.html

PART IV

Spelling and Mechanics

CHECKLIST

Spelling and Mechanics

This checklist covers the main conventions of spelling and mechanics. For a detailed guide to this part, see the contents inside the back cover.

Spelling

☑ Proofread for correct spelling. Don't rely on your spelling checker. (See pp. 91–94.)

Capital letters

☑ Use capital letters appropriately for proper nouns and adjectives and for the titles of works and persons. (See pp. 94–96.)

Underlining or italics

☑ Use underlining or italics primarily for the titles of works published separately from other works. Ensure that underlining or italics in source citations conforms to your discipline's or instructor's requirements. (See pp. 96–98.)

Abbreviations

☑ Use abbreviations appropriately for the discipline or field you are writing in. (See pp. 99–100.)

Numbers

☑ Express numbers in numerals or words appropriately for the discipline or field you are writing in. (See pp. 100–01.)

26 Spelling and the Hyphen

Spelling, including using the hyphen, is a skill you can acquire by paying attention to words and by developing three habits:

- Carefully proofread your writing.
- Cultivate a healthy suspicion of your spellings.
- Habitually check a dictionary when you doubt a spelling.

26a ▪ Spelling checkers

A spelling checker can help you find and track spelling errors in your papers. But its usefulness is limited, mainly because it can't spot the confusion of words with similar spellings, such as *their/they're/there*. A grammar and style checker may flag such words, but only the ones listed in its dictionary, and you still must select the correct spelling. Proofread your papers to catch spelling errors.

See page 7 for more on spelling checkers.

26b

26b ▪ Spelling rules

We often misspell syllables rather than whole words. The following rules focus on troublesome syllables.

ie and *ei*

Follow the familiar jingle: *i* before *e* except after *c* or when pronounced "ay" as in *neighbor* and *weigh*.

i before *e*	believe	thief
e before *i*	receive	ceiling
ei pronounced "ay"	freight	vein

Exceptions Remember common exceptions with this sentence: *The weird foreigner neither seizes leisure nor forfeits height.*

Silent final *e*

Drop a silent *e* when adding an ending that begins with a vowel. Keep the *e* if the ending begins with a consonant.

advise + able = advisable   care + ful = careful

Visit *www.ablongman.com/littlebrown* for added help and exercises on spelling and the hyphen.

Exceptions Keep the final *e* before a vowel to prevent confusion or mispronunciation: *dyeing, changeable.* Drop the *e* before a consonant when another vowel comes before the *e: argument, truly.*

Final *y*

When adding to a word ending in *y,* change *y* to *i* when it follows a consonant. Keep the *y* when it follows a vowel, precedes *-ing,* or ends a proper name.

beauty, beauties	day, days
worry, worried	study, studying
supply, supplier	Minsky, Minskys

Consonants

When adding an ending to a one-syllable word that ends in a consonant, double the consonant if it follows a single vowel. Otherwise, don't double the consonant.

slap, slapping	pair, paired

For words of more than one syllable, double the final consonant when it follows a single vowel and ends a stressed syllable once the new ending is added. Otherwise, don't double the consonant.

submit, submitted	despair, despairing
refer, referred	refer, reference

Plurals

Most nouns form plurals by adding *s* to the singular form. Nouns ending in *s, sh, ch,* or *x* add *es* to the singular.

boy, boys	kiss, kisses
table, tables	lurch, lurches
Murphy, Murphys	tax, taxes

Nouns ending in a vowel plus *o* add *s.* Nouns ending in a consonant plus *o* add *es.*

ratio, ratios	hero, heroes

Form the plural of a compound noun by adding *s* to the main word in the compound. The main word may not fall at the end.

passersby	breakthroughs
fathers-in-law	city-states

Some English nouns that come from other languages form the plural according to their original language.

analysis, analyses	medium, media
crisis, crises	phenomenon, phenomena
criterion, criteria	piano, pianos

ESL Noncount nouns do not form plurals, either regularly (with an added *s*) or irregularly. Examples of noncount nouns include *equipment, intelligence,* and *wealth.* See pages 57–58.

American vs. British or Canadian spellings ESL

American and British or Canadian spellings differ in ways such as the following.

American	British or Canadian
color, humor	colour, humour
theater, center	theatre, centre
canceled, traveled	cancelled, travelled
judgment	judgement
realize, analyze	realise, analyse
defense, offense	defence, offence

26c ▪ The hyphen

Use the hyphen to form compound words and to divide words at the ends of lines.

Compound words

26c

Compound words may be written as a single word (*breakthrough*), as two words (*decision makers*), or as a hyphenated word (*cross-reference*). Check a dictionary for the spelling of a compound word. Except as explained below, any compound not listed in the dictionary should be written as two words.

Sometimes a compound word comes from combining two or more words into a single adjective.° When such a compound adjective precedes a noun, a hyphen forms the words clearly into a unit.

She is a well-known actor.
Some Spanish-speaking students work as translators.

When a compound adjective follows the noun, the hyphen is unnecessary.

The actor is well known.
Many students are Spanish speaking.

The hyphen is also unnecessary in a compound modifier containing an *-ly* adverb, even before the noun: *clearly defined terms.*

Fractions and compound numbers

Hyphens join the parts of fractions: *three-fourths, one-half.* And the whole numbers *twenty-one* to *ninety-nine* are always hyphenated.

Prefixes and suffixes

Do not use hyphens with prefixes except as follows:

- With the prefixes *self-*, *all-*, and *ex-*: *self-control, all-inclusive, ex-student.*
- With a prefix before a capitalized word: *un-American.*
- With a capital letter before a word: *T-shirt.*
- To prevent misreading: *de-emphasize, re-create a story.*

The only suffix that regularly requires a hyphen is *-elect,* as in *president-elect.*

Word division at the end of a line

You can avoid occasional short lines in your documents by setting your word processor to divide words automatically at appropriate breaks. To divide words manually, follow these guidelines:

- Divide words only between syllables—for instance, *win-dows,* not *wi-ndows.* Check a dictionary for correct syllable breaks.
- Never divide a one-syllable word.
- Leave at least two letters on the first line and three on the second line. If a word cannot be divided to follow this rule (for instance, *a-bus-er*), don't divide it.

If you must break an electronic address—for instance, in a source citation—do so only after a slash. Do not hyphenate, because readers may perceive any added hyphen as part of the address.

Not http://www.library.miami.edu/staff/lmc/soc-
race.html

But http://www.library.miami.edu/staff/lmc/
socrace.html

27 Capital Letters

As a rule, capitalize a word only when a dictionary or conventional use says you must. Consult one of the style guides listed on pages 141–42 for special uses of capitals in the social, natural, and applied sciences.

Visit *www.ablongman.com/littlebrown* for added help and an exercise on capital letters.

Note A grammar and style checker will flag overused capital letters and missing capitals at the beginnings of sentences. It will also spot missing capitals at the beginnings of proper nouns and adjectives—*if* the nouns and adjectives are in the checker's dictionary. For example, a checker caught *christianity* and *europe* but not *china* (for the country) or *Stephen king*. You'll need to proofread for capital letters on your own as well.

27a ▪ First word of a sentence

Every writer should own a good dictionary.

27b ▪ Proper nouns° and proper adjectives°

Specific persons and things

Stephen King	Boulder Dam

Specific places and geographical regions

New York City	the Northeast, the South

But: northeast of the city, going south

Days of the week, months, holidays

Monday	Yom Kippur
May	Christmas

But: winter, spring, summer, fall

Historical events, documents, periods, movements

the Vietnam War	the Renaissance
the Constitution	the Romantic Movement

Government offices or departments and institutions

Polk Municipal Court	House of Representatives
Northeast High School	Department of Defense

Organizations, associations, and their members

B'nai B'rith	Democratic Party, Democrats
Rotary Club	League of Women Voters
Atlanta Falcons	Chicago Symphony Orchestra

Races, nationalities, and their languages

Native American	Germans
African American	Swahili
Caucasian	Italian

But: blacks, whites

Religions, their followers, and terms for the sacred

Christianity, Christians	God
Catholicism, Catholics	Allah
Judaism, Orthodox Jew	the Bible (*but* biblical)
Islam, Muslims	the Koran, the Qur'an

Common nouns as parts of proper nouns

Main Street	Lake Superior
Central Park	Ford Motor Company
Pacific Ocean	Madison College

But: the ocean, college course, the company

27c ▪ Titles and subtitles of works

Capitalize all the words in a title and subtitle *except* articles (*a, an, the*), *to* in infinitives,° and connecting words of fewer than five letters. Capitalize even these words when they are the first or last word in the title or when they fall after a colon or semicolon.

"Once More to the Lake"	*Management: A New Theory*
Learning from Las Vegas	"The Truth About AIDS"
"Knowing Whom to Ask"	*An End to Live For*

Note Some academic disciplines require other treatments of titles within source citations, such as capitalizing different words or only the first words. See pages 147 (MLA style), 166 (APA style), and 193 (CSE style).

27d ▪ Titles of persons

Title before name	Professor Jane Covington
Title after name	Jane Covington, a professor

27e ▪ Online communication

Online messages written in all-capital letters or with no capitals are difficult to read, and those in all-capitals are often considered rude. Use capital letters according to the rules in 27a–27c in all your online communication.

28 Underlining or Italics

Underlining and *italic type* indicate the same thing: the word or words are being distinguished or emphasized.

Note Grammar and style checkers cannot recognize problems with underlining or italics. Check your work yourself to ensure that any highlighting is appropriate.

Visit *www.ablongman.com/littlebrown* for added help and an exercise on underlining or italics.

28a ▪ Underlining vs. italics

Italic type is now used almost universally in business and academic writing. Still, some academic style guides, notably the *MLA Handbook,* continue to prefer underlining, especially in source citations. Ask your instructor for his or her preference. (Underlining is used for the examples in this chapter because it is easier to see than italics.)

Use either italics or underlining consistently throughout a document. For instance, if you are writing an English paper and following MLA style for underlining in source citations, use underlining in the body of your paper as well.

28b ▪ Titles of works

Do not underline or italicize the title of your own paper unless it contains an element (such as a book title) that requires highlighting.

Within your text, underline or italicize the titles of works that are published, released, or produced separately from other works. Use quotation marks for all other titles (see p. 81).

Book
War and Peace

Play
Hamlet

Web site
Friends of Prufrock
Google

Computer software
Microsoft Internet Explorer
Acrobat Reader

Long musical work
The Beatles' *Revolver*
But: Symphony in C

Work of visual art
Michelangelo's *David*

Long poem
Paradise Lost

Periodical
Philadelphia Inquirer

Television or radio program
60 Minutes

Movie
Psycho

Pamphlet
The Truth About Alcoholism

Published speech
Lincoln's *Gettysburg Address*

Exceptions Legal documents, the Bible, and parts of them are generally not underlined or italicized.

We studied the Book of Revelation in the Bible.

28c ▪ Ships, aircraft, spacecraft, trains

Challenger
Apollo XI
Orient Express
Montrealer
Queen Elizabeth 2
Spirit of St. Louis

28d ▪ Foreign words and phrases

The scientific name for the brown trout is Salmo trutta. [The Latin scientific names for plants and animals are always highlighted.]

The Latin De gustibus non est disputandum translates roughly as "There's no accounting for taste."

28e ▪ Words or characters named as words

Underline or italicize words or characters (letters or numbers) that are referred to as themselves rather than used for their meanings. Such words include terms you are defining.

Some children pronounce th, as in thought, with an f sound.

The word syzygy refers to a straight line formed by three celestial bodies, as in the alignment of the earth, sun, and moon.

28f ▪ Online alternatives

Electronic mail and other forms of online communication may not allow conventional highlighting such as underlining or italics for the purposes described above. The program may not be able to produce the highlighting or may reserve it for a special function. (On Web sites, for instance, underlining often indicates a link to another site.)

To distinguish book titles and other elements that usually require underlining or italics, type an underscore before and after the element: *Measurements coincide with those in _Joule's Handbook_.* You can also emphasize words with asterisks before and after: _I *will not* be able to attend._

Don't use all-capital letters for emphasis; they yell too loudly. (See also p. 96.)

29 Abbreviations

The following guidelines on abbreviations pertain to the text of a nontechnical document. For any special requirements of the discipline you are writing in, consult one of the style guides listed on pages 141–42. In all disciplines, writers increasingly omit periods from abbreviations that end in capital letters. (See p. 83.)

If a name or term (such as *operating room*) appears often in a piece of writing, then its abbreviation (*OR*) can cut down on extra words. Spell out the full term at its first appearance, give the abbreviation in parentheses, and use the abbreviation thereafter.

Note A grammar and style checker may flag some abbreviations such as *ft.* (for *foot*) and *st.* (for *street*). A spelling checker will flag abbreviations it does not recognize. But neither checker can tell you whether an abbreviation is appropriate for your writing situation or will be clear to your readers.

29a ▪ Familiar abbreviations

Titles before names	Dr., Mr., Mrs., Ms., Rev., Gen.
Titles after names	MD, DDS, DVM, PhD, Sr., Jr.
Institutions	LSU, UCLA, TCU, NASA
Organizations	CIA, FBI, YMCA, AFL-CIO
Corporations	IBM, CBS, ITT, GM
People	JFK, LBJ, FDR
Countries	USA, UK
Specific numbers	no. 36 *or* No. 36
Specific amounts	$7.41, $1 million
Specific times	11:26 AM, 8:04 a.m., 2:00 PM, 8:05 p.m.
Specific dates	44 BC, AD 1492, 44 BCE, 1492 CE

Note The abbreviation AD (*anno domini,* "in the year of the Lord") always precedes the date. Increasingly, AD and BC are being replaced by CE ("common era") and BCE ("before the common era").

Visit ***www.ablongman.com/littlebrown*** for added help and an exercise on abbreviations.

29b ▪ Latin abbreviations

Generally, use the common Latin abbreviations (without italics or underlining) only in source citations and comments in parentheses.

i.e. *id est:* that is
cf. *confer:* compare
e.g. *exempli gratia:* for example
et al. *et alii:* and others
etc. *et cetera:* and so forth
NB *nota bene:* note well

He said he would be gone a fortnight (i.e., two weeks).
Bloom et al., editors, *Anthology of Light Verse*

29c ▪ Words usually spelled out

Generally spell out certain kinds of words in the text of academic, general, and business writing. (In technical writing, however, abbreviate units of measurement.)

Units of measurement

Mount Everest is 29,028 feet high.

Geographical names

Lincoln was born in Illinois.

Names of days, months, and holidays

The truce was signed on Tuesday, April 16, and was ratified by Christmas.

Names of people

Robert Frost wrote accessible poems.

Courses of instruction

The writer teaches psychology and composition.

And

The new rules affect New York City and environs. [Use the ampersand, &, only in the names of business firms: *Lee & Sons*.]

30 Numbers

This chapter addresses the use of numbers (numerals versus words) in the text of a nontechnical document. Usage

Visit ***www.ablongman.com/littlebrown*** for added help and an exercise on numbers.

does vary, so consult one of the style guides listed on pages 141–42 for the requirements of the discipline you are writing in.

Note Grammar and style checkers will flag numerals beginning sentences and can be customized to ignore or to look for numerals. But they can't tell you whether numerals or spelled-out numbers are appropriate for your writing situation.

30a ▪ Numerals, not words

Numbers requiring three or more words

366 36,500

Round numbers over a million

26 million 2.45 billion

Exact amounts of money

$3.5 million $4.50

Days and years

June 18, 1985 AD 12
456 BC 12 CE

The time of day

9:00 AM 3:45 PM

Addresses

355 Heckler Avenue
Washington, DC 20036

Decimals, percentages, and fractions

22.5 3½
48% (*or* 48 percent)

Scores and statistics

a ratio of 8 to 1 21 to 7
a mean of 26

Pages, chapters, volumes, acts, scenes, lines

Chapter 9, page 123
Hamlet, Act 5, Scene 3

30b ▪ Words, not numerals

Numbers of one or two words

sixty days, forty-two laps, one hundred people

In business and technical writing, use words only for numbers under 11 (*ten reasons, four laps*).

Beginnings of sentences

Seventy-five riders entered the race.

If a number beginning a sentence requires more than two words to spell out, reword the sentence so that the number falls later.

Faulty 103 visitors returned.
Awkward One hundred three visitors returned.
Revised Of the visitors, 103 returned.

PART V

Research and Documentation

CHECKLIST

Research and Documentation

This checklist covers the main considerations in using and documenting sources. For a detailed guide to this part, see the contents inside the back cover.

Developing a research strategy

- ☑ Formulate a question about your subject that can guide your research. (See opposite.)
- ☑ Set goals for your sources: library vs. Internet, primary vs. secondary, and so on. (See opposite.)
- ☑ Anticipate the aims and methods of gathering source information. (See p. 108.)
- ☑ Prepare a working bibliography to keep track of sources. (See p. 109.)

Finding sources

- ☑ Develop keywords that describe your subject for searches of electronic sources. (See p. 112.)
- ☑ Consult appropriate sources to answer your research question. (See p. 113.)

Evaluating and synthesizing sources

- ☑ Evaluate both print and online sources for their relevance and reliability. (See p. 122.)
- ☑ Synthesize sources to find their relationships and to support your own ideas. (See p. 126.)

Integrating sources into your text

- ☑ Summarize, paraphrase, or quote sources depending on the significance of the source's ideas or wording. (See p. 128.)
- ☑ Work source material smoothly and informatively into your own text. (See p. 130.)

Avoiding plagiarism and documenting sources

- ☑ Do not plagiarize, either deliberately or accidentally, by presenting the words or ideas of others as your own. (See p. 136.)
- ☑ Using the style guide appropriate for your discipline, document your sources and format your paper. (See pp. 141–42 for a list of style guides and Chapters 36 for MLA style, 37 for APA style, 38 for Chicago style, and 39 for CSE style.)

31 Developing a Research Strategy

Research writing gives you a chance to work like a detective solving a case. The mystery is the answer to a question you care about. The search for the answer leads you to consider what others think about your subject, to build on that information, and ultimately to become an expert in your own right.

31a ▪ Subject, question, and thesis

Seek a research subject that you care about and want to know more about. Starting with such an interest and with your own views will motivate you and will make you a participant in a dialog when you begin examining sources.

Asking a question about your subject can give direction to your research by focusing your thinking on a particular approach. To discover your question, consider what about your subject intrigues or perplexes you, what you'd like to know more about. (See the next page for suggestions on using your own knowledge.)

Try to narrow your research question so that you can answer it in the time and space you have available. The question *How does the Internet affect business?* is very broad, encompassing issues as diverse as electronic commerce, information management, and employee training. In contrast, the question *How does Internet commerce benefit consumers?* or *How, if at all, should Internet commerce be taxed?* is much narrower. Each question also requires more than a simple yes-or-no answer, so that answering, even tentatively, demands thought about pros and cons, causes and effects.

As you read and write, your question will probably evolve to reflect your increasing knowledge of the subject. Eventually its answer will become the **thesis** of your paper, the main idea that all the paper's evidence supports. (See also pp. 3–4.)

31b ▪ Goals for sources

Before you start looking for sources, consider what you already know about your subject and where you are likely to find information on it.

Visit *www.ablongman.com/littlebrown* for added help and an exercise on research strategy.

Your own knowledge

Discovering what you already know about your topic will guide you in discovering what you don't know and thus need to research. Take some time to spell out facts you have learned, opinions you have heard or read elsewhere, and of course your own opinions.

When you've explored your thoughts, make a list of questions for which you don't have answers, whether factual (*What laws govern taxes in Internet commerce?*) or more open-ended (*Who benefits from a tax-free Internet? Who doesn't benefit?*). These questions will give you clues about the sources you need to look for first.

Kinds of sources

For many research projects, you'll want to consult a mix of sources, as described below. You may start by seeking the outlines of your subject—the range and depth of opinions about it—in reference works and articles in popular periodicals or through a search of the Web. Then, as you refine your views and your research question, you'll move on to more specialized sources, such as scholarly books and periodicals and your own interviews or surveys. (See pp. 113–21 for more on each kind of source.)

Library and Internet sources

The print and electronic sources available through your library—mainly reference works, periodicals, and books—have two big advantages over most of what you'll find on the Internet: they are cataloged and indexed for easy retrieval; and they are generally reliable, having been screened first by their publishers and then by the library's staff. In contrast, the Internet's retrieval systems are more difficult to use effectively, and Web sources can be less reliable because many do not pass through any screening before being posted.

Most instructors expect research writers to consult library sources. But they'll accept online sources, too, if you have used them judiciously. Even with its disadvantages, the Internet can be a valuable resource for primary sources, scholarly work, current information, and diverse views. For guidelines on evaluating both print and online sources, see pages 122–26.

Primary and secondary sources

As much as possible, you should rely on **primary sources**. These are firsthand accounts, such as historical documents (letters, speeches, and so on), eyewitness reports, works of literature, reports on experiments or surveys conducted by the writer, and your own interviews, experiments, observations, or correspondence.

In contrast, **secondary sources** report and analyze information drawn from other sources, often primary ones: a reporter's summary of a controversial issue, a historian's account of a battle, a critic's reading of a poem, a psychologist's evaluation of several studies. Secondary sources may contain helpful summaries and interpretations that direct, support, and extend your own thinking. However, most research-writing assignments expect your own ideas to go beyond those in such sources.

Scholarly and popular sources

The scholarship of acknowledged experts is essential for depth, authority, and specificity. The general-interest views and information of popular sources can help you apply more scholarly approaches to daily life.

When looking for sources, you can gauge how scholarly or popular they are from bibliographic information:

- *Check the publisher.* Is it a scholarly journal (such as *Education Forum*) or a publisher of scholarly books (such as Harvard University Press), or is it a popular magazine (such as *Time* or *Newsweek*) or a publisher of popular books (such as Little, Brown)?
- *Check the author.* Have you seen the name elsewhere, which might suggest that the author is an expert?
- *Check the title.* Is it technical, or does it use a general vocabulary?
- *Check the URL.* Addresses for Internet sources often include an abbreviation that tells you something about the source: *edu* means the source comes from an educational institution, *gov* from a government body, *org* from a nonprofit organization, *com* from a commercial organization such as a corporation. (See p. 124 for more on interpreting electronic addresses.)

Older and newer sources

For most subjects a combination of older, established sources (such as books) and current sources (such as newspaper articles or interviews) will provide both background and up-to-date information. Only historical subjects or very current subjects require an emphasis on one extreme or another.

Impartial and biased sources

Seek a range of viewpoints. Sources that attempt to be impartial can offer an overview of your subject and trustworthy facts. Sources with clear biases can offer a diversity of opinion. Of course, to discover bias, you may have to read the source carefully (see p. 123); but even a bibliographical listing can be informative.

- *Check the author.* You may have heard of the author as a respected researcher (thus more likely to be objective) or as a leading proponent of a certain view (less likely to be objective).
- *Check the title.* It may reveal something about point of view. (Consider these contrasting titles: "Keep the Internet Tax-Free" and "Taxation of Internet Commerce: Issues and Questions.")

Note Internet sources must be approached with particular care. See pages 123–26.

Sources with helpful features

Depending on your topic and how far along your research is, you may want to look for sources with features such as illustrations (which can clarify important concepts), bibliographies (which can direct you to other sources), and indexes (which can help you develop keywords for electronic searches; see p. 112).

31c ▪ Information gathering

All methods for gathering source information share the same goals:

- *Keep accurate records of what sources say.* Accuracy helps prevent misrepresentation and plagiarism.
- *Keep accurate records of how to find sources.* These records are essential for retracing steps and for citing sources in the final paper. (See opposite on keeping a working bibliography.)
- *Interact with sources.* Reading sources critically leads to an understanding of them, the relationships among them, and their support for one's own ideas.

Each method of gathering information has advantages and disadvantages. On any given project, you may use all the methods.

- *Handwritten notes:* Taking notes by hand is especially useful if you come across a source with no computer or photocopier handy. But handwritten notes can be risky. It's easy to introduce errors as you work from source to note card. And it's possible to copy source language and then later mistake and use it as your own, thus plagiarizing the source. Always take care to make accurate notes and to place big quotation marks around any passage you quote.

- *Notes on computer:* Taking notes on your computer can streamline the path of source to note to paper, because you can import the notes into your draft as you write. However, computer notes share the disadvantages of handwritten notes: the risks of introducing errors and of plagiarizing. As with handwritten notes, strive for accuracy, and use quotation marks for quotations.
- *Photocopies and printouts:* Photocopying from print sources or printing out online sources is convenient and reduces the risks of error and plagiarism. But both methods have disadvantages, too. The busywork of copying or printing can distract you from the crucial work of interacting with sources. And you have to make a special effort to annotate copies and printouts with the publication information for sources. If you don't have this information for your final paper, you can't use the source.
- *Downloads:* Researching online, you can usually download full-text articles, Web pages, and other materials into your word processor and then into your drafts. Like photocopies and printouts, though, downloads can distract you from interacting with sources and can easily become separated from the publication information you must have in order to use the sources. Even more important, directly importing source material creates a high risk of plagiarism. You must keep clear boundaries between your own ideas and words and those of others.

31d ▪ Working bibliography

Keep records of sources so that you can always locate them and can fully acknowledge them in your paper. Use the following lists to track different kinds of sources. For online sources, you may need extra effort to uncover the required information (see the next page).

Bibliographic information

For a book

Library call number
Name(s) of author(s), editor(s), translator(s), or others listed
Title and subtitle
Publication data: (1) place of publication; (2) publisher's name; (3) date of publication
Other important data, such as edition or volume number

For a periodical article

Name(s) of author(s)
Title and subtitle of article
Title of periodical

Publication data: (1) volume number and issue number (if any) in which the article appears; (2) date of issue; (3) page numbers on which article appears

For electronic sources

Name(s) of author(s)
Title and subtitle
Publication data for books and articles (see above)
Date of release, online posting, or latest revision
Medium (online, CD-ROM, etc.)
Format of online source (Web site, Web page, e-mail, etc.)
Date you consulted the source
Complete URL or electronic address (unless source was obtained through a subscription service and has no permanent address)
For sources obtained through a subscription service: (1) name of database; (2) name of service; (3) electronic address of the service's home page *or* search terms used to reach the source

For other sources

Name(s) of author(s) or others listed, such as a government department or a recording artist
Title of the work
Format, such as unpublished letter or live performance
Publication or production data: (1) publisher's or producer's name; (2) date of publication, release, or production; (3) identifying numbers (if any)

Online sources

The following guidelines can help you find and record the publication information for online sources, such as the Web page in the screen shot on the facing page.

1. *Find the source's URL, or address, in the Web browser's Address or Location field near the top of the screen.* Generally, you'll use this URL in acknowledging the source. With a source you find through a library subscription service, however, the URL may be unusable. See pages 155–56 for what to do in that case.
2. *Use as the source title the title of the page you are consulting.* This information usually appears as a heading at the top of the page, but if not it may also appear in the bar along the top of the browser window.
3. *Look for publication information at the bottom of each page.* Check here for (*a*) the name of the author or the sponsoring organization, (*b*) an address for reaching the sponsor or author directly, and (*c*) the date of publication or last revision.

If the page you are reading does not list publication information, look for it on the site's home page. There

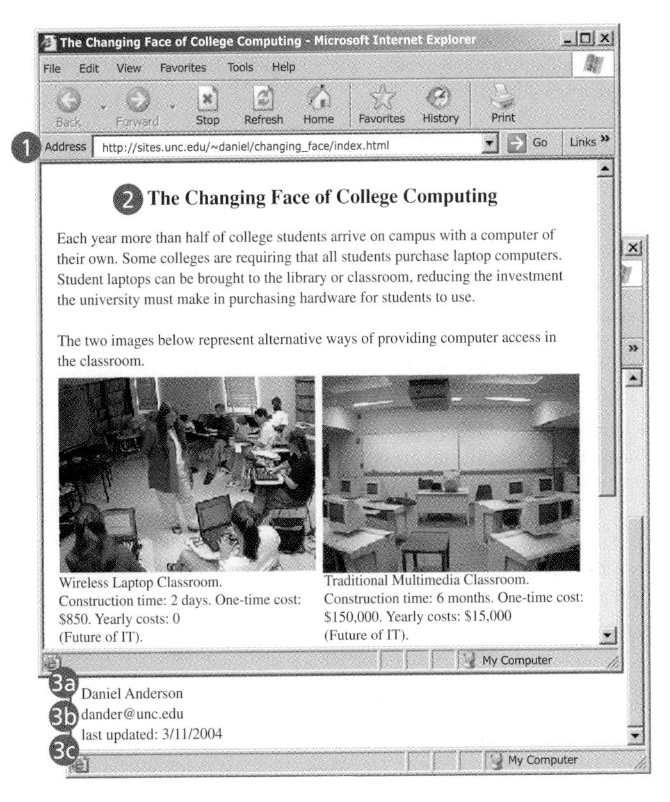

may be a link to the home page, or you can find it by editing the URL in the Address or Location field: working backward, delete the end of the URL up to the preceding slash; then hit Enter. (For the address in the screen shot, you would delete *index.html*.) If that doesn't take you to the home page, delete the end of the remaining address up to the preceding slash and hit Enter. Editing the address in this way, you'll eventually reach the home page.

32 Finding Sources

Your library and a computer connected to the Internet give you access to a vast range of sources. This chapter explains the basics of an electronic search and introduces

Visit *www.ablongman.com/littlebrown* for added help and an exercise on finding sources.

reference works, books, periodicals, the Web, online-communication sources, government publications, and your own sources.

32a ▪ Electronic searches

As you conduct research, the World Wide Web will be your gateway to ideas and information. Always begin your academic research on your library's Web site, not with a search engine such as *Google*. Although a direct Web search can be productive, you'll find a higher concentration of relevant and reliable sources through the library.

Probably the most important element in an electronic search is appropriate **keywords,** or **descriptors,** that name your subject for databases and Web search engines. With a database, such as a periodical index, your keywords will help you find the subject headings used by the database to organize the sources you seek. With a Web search engine, such as *Google,* your keywords must match terms used in source titles and texts. In either case, you may have to use trial and error to discover keywords that accurately describe your subject.

Most databases and search engines work with **Boolean operators,** terms or symbols that allow you to expand or limit your keywords and thus your search. The basic operators appear below, but resources do differ. The Help information of each resource explains its system.

- *Use* AND *or + to narrow the search* by including only sources that use all the given words. The keywords *Internet AND tax* request only the sources in the shaded area:

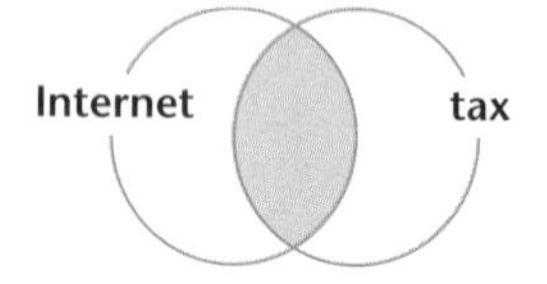

- *Use* NOT *or – ("minus") to narrow the search* by excluding irrelevant words. *Internet AND tax NOT access* excludes sources that use the word *access.*

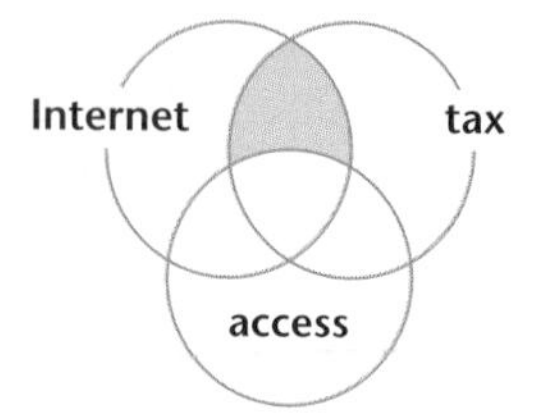

- *Use* OR *to broaden the search* by giving alternative keywords. *Internet OR (electronic commerce) AND tax* allows for sources that use *Internet* or *electronic commerce* (or both) along with *tax.*

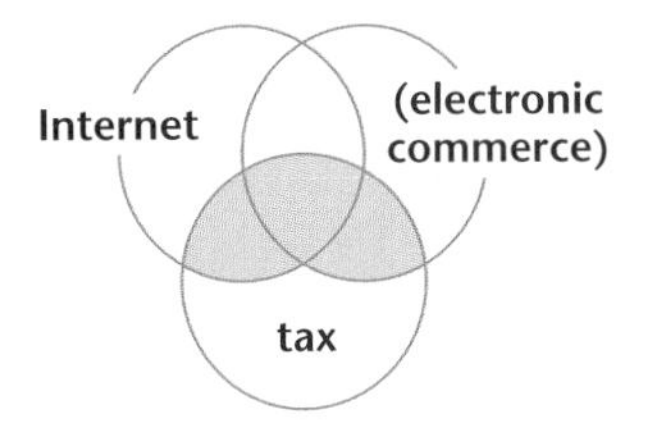

- *Use parentheses to form search phrases.* For instance, *(electronic commerce)* requests the exact phrase, not the separate words. (Some systems use quotation marks instead of parentheses for this purpose.)
- *Use* NEAR *to narrow the search* by requiring the keywords to be close to each other—for instance, *Internet NEAR tax.* Depending on the resource you're using, the words could be directly next to each other or many words apart. Some resources use *WITHIN* __ so that you can specify the exact number of words apart—for instance, *Internet WITHIN 10 tax.*
- *Use wild cards to permit different versions of the same word.* In *child**, for instance, the wild card * indicates that sources may include *child, children, childcare, childhood, childish, childlike,* and *childproof.* If a wild card opens up your search too much (as with the example *child**), you may be better off using OR to limit the options: *child OR children.* (Some systems use ?, :, or + for a wild card instead of *.)
- *Be sure to spell your keywords correctly.* Some search tools will look for close matches or approximations, but correct spelling gives you the best chance of finding relevant sources.

See pages 118–20 for a sample keyword search of the Web.

32b

32b ▪ Reference works

Reference works (often available online) include encyclopedias, dictionaries, digests, bibliographies, indexes, handbooks, atlases, and almanacs. Your research *must* go beyond these sources, but they can help you decide whether your subject really interests you, can help you develop your keywords for electronic searches, and can direct you to more detailed sources.

You'll find many reference works through your library and directly on the Web. The following lists give general Web references for academic research. Discipline-specific references and updates of those below appear on this book's Web site at *www.ablongman.com/littlebrown.*

All disciplines

Internet Public Library
http://www.ipl.org
Library of Congress
http://lcweb.loc.gov
LSU Libraries Webliography
http://www.lib.lsu.edu/weblio.html
World Wide Web Virtual Library
http://vlib.org

Humanities

BUBL Humanities Resources
http://bubl.ac.uk/link
EDSITEment
http://edsitement.neh.gov
Voice of the Shuttle Humanities Gateway
http://vos.ucsb.edu

Social sciences

Social Science Data on the Internet
http://odwin.ucsd.edu/idata
Social Science Information Gateway
http://www.sosig.ac.uk
Virtual Library: Social Sciences
http://vlib.org

Natural and applied sciences

Google Web Directory Science Links
http://directory.google.com/Top/Science
Librarians' Index Science Resources
http://lii.org/search/file/scitech
Virtual Library: Science
http://vlib.org

32c ▪ Books

Depending on your subject, you may find some direction and information in so-called general or trade books, which include personal or popular views of culture, nonspecialist explanations of scholarly work, and how-to guides. But usually the books you consult for academic research will be scholarly, intended for specialists, and will include detailed statements of theory or surveys of research.

You can search your library's online catalog by authors' names, titles, or keywords. Using the tips on pages

112–13, experiment with your own keywords until you locate a source that seems relevant to your subject. The book's detailed record will show Library of Congress subject headings that you can use to find similar books.

32d ▪ Periodicals

Periodicals include newspapers, journals, and magazines. Newspapers, the easiest to recognize, are useful for detailed accounts of past and current events. Journals and magazines can be harder to distinguish, but their differences are important. Most college instructors expect students' research to rely more on journals than on magazines.

Journals	Magazines
Examples: *American Anthropologist, Journal of Black Studies, Journal of Chemical Education*	Examples: *The New Yorker, Time, Rolling Stone, People*
Articles are intended to advance knowledge in a particular field.	Articles are intended to express opinion, inform, or entertain.
Writers and readers are specialists in the field.	Writers may or may not be specialists in their subjects. Readers are members of the general public or a subgroup with a particular interest.
Articles always include source citations.	Articles rarely include source citations.
Appearance is bland, with black-only type, little or no decoration, and only illustrations that directly amplify the text, such as graphs.	Appearance varies but is generally lively, with color, decoration (headings, sidebars, and other elements), and illustrations (drawings, photographs).
Issues may appear quarterly or less often.	Issues may appear weekly, biweekly, or monthly.
Issues may be paged separately (like a magazine) or may be paged sequentially throughout an annual volume, so that issue 3 (the third issue of the year) could open on page 327. (The method of pagination affects source citations. See p. 152.)	Issues are paged separately, each beginning on page 1.

Periodical articles are cataloged in printed or electronic indexes. Your library subscribes to many indexes and to multi-index databases, such as the following:

EBSCOhost Academic Search. A periodical index covering magazines and journals in the social sciences, sciences, arts, and humanities. Many articles are available in full text.

InfoTrac Expanded Academic. The Gale Group's general periodical index covering the social sciences, sciences, arts, and humanities as well as national news periodicals. It includes full-text articles.

LexisNexis Academic. An index of news and business, legal, and reference information, with full-text articles. *LexisNexis* includes international, national, and regional newspapers, newsmagazines, and legal and business publications.

Nineteenth-Century Masterfile. Perhaps the only electronic database for periodicals from the nineteenth century.

ProQuest Research Library. A periodical index covering the sciences, social sciences, arts, and humanities, including many full-text articles.

Wilson Databases. A collection of indexes, often provided in a package, including *Business Periodicals Index, Education Index, General Science Index, Humanities Index, Readers' Guide to Periodical Literature,* and *Social Sciences Index*.

Check the library's list of indexes to find the ones that seem most appropriate for your subject. (The library's Web site may include a search function to help you with this step.) Search the indexes themselves as discussed on pages 112–13.

The screen shot on the facing page shows partial results from a search of *EBSCOhost Academic Search*. Many items are available online in full text. The others are available in the library or through interlibrary loan. Clicking on any of these items would lead to a more detailed record and an **abstract**, or summary, of the article. An abstract can tell you whether you want to pursue an article further. It is not the article, however, and should not be used or cited as if it were.

Note Many databases allow you to limit your search to so-called peer-reviewed or refereed journals—that is, scholarly journals whose articles have been reviewed before publication by experts in the field and then revised by the author. Limiting your search to peer-reviewed journals can help you navigate huge databases that might otherwise return scores of unusable articles.

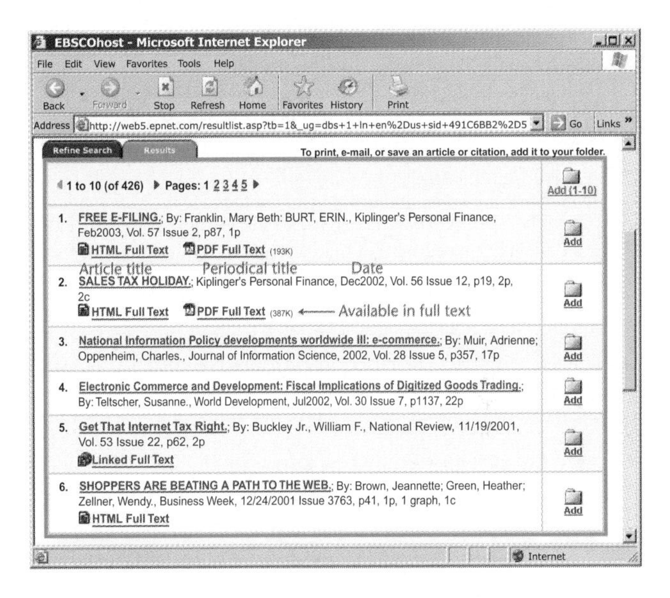

32e ▪ The Web

As an academic researcher, you enter the World Wide Web in two ways: through your library's Web site, and through public search engines such as *Yahoo!* and *Google*. The library entrance, covered in the preceding sections, is your main path to the books and periodicals that, for most subjects, should make up most of your sources. The public entrance, discussed here, can lead to a wealth of information and ideas, but it also has a number of disadvantages:

- *The Web is a wide-open network.* Anyone with the right hardware and software can place information on the Internet, and even a carefully conceived search can turn up sources with widely varying reliability: journal articles, government documents, scholarly data, term papers written by high school students, sales pitches masked as objective reports, wild theories. You must be especially diligent about evaluating Internet sources (see p. 123).
- *The Web changes constantly.* No search engine can keep up with the Web's daily additions and deletions, and a source you find today may be different or gone tomorrow. Some sites are designed and labeled as archives: they do not change except with additions. But generally you should not put off consulting an online source that you think you may want to use.

- *The Web provides limited information on the past.* Sources dating from before the 1980s or even more recently probably will not appear on the Web.
- *The Web is not all-inclusive.* Most books and many periodicals are available only via the library, not directly via the Web.

Clearly, the Web warrants cautious use. It should not be the only resource you work with.

Current search engines

The following lists include the currently most popular search engines. For a comprehensive and up-to-date guide to the latest search engines, consult *Search Engine Watch* at *http://searchenginewatch.com/links.*

- Several directories review sites:

  BUBL Link (*http://bubl.ac.uk/link*)
  Internet Public Library (*http://www.ipl.org/div/subject*)
  Internet Scout Project (*http://scout.wisc.edu/archives*)
  Librarians' Index to the Internet (*http://lii.org*)

- Two search engines are considered the most advanced and efficient:

  AlltheWeb (*http://www.alltheweb.com*): One of the fastest and most comprehensive engines, *AlltheWeb* updates its database frequently so that it returns more of the Web's most recent sites.

  Google (*http://www.google.com*): Also fast and comprehensive, *Google* ranks a site based not only on its content but also on the other sites that are linked to it, thus providing a measure of a site's usefulness.

- A number of other engines can back up *AlltheWeb* and *Google:*

  AltaVista (*http://www.altavista.com*)
  Ask Jeeves (*http://www.ask.com*)
  Dogpile (*http://www.dogpile.com*)
  Excite (*http://www.excite.com*)
  Lycos (*http://www.lycos.com*)
  MetaCrawler (*http://www.metacrawler.com*)
  Yahoo! (*http://www.yahoo.com*)

A sample search

The sample Web search illustrated on the facing page shows how the refinement of keywords can narrow a search to maximize the relevant hits and minimize the irrelevant ones. Kisha Alder, a student researching the feasibility of Internet sales taxes, first used the keywords *taxes AND Internet* on *Google* (first screen shot). But the search produced more than 2 million hits, an unusably large number and a sure sign that Alder's keywords needed revision.

After several tries, Alder arrived at two phrases, *sales tax* and *electronic commerce,* to describe her subject more precisely. Taking advantage of a feature described on *Google*'s Advanced Search help, she added *site:.gov* at the end

1. First *Google* results

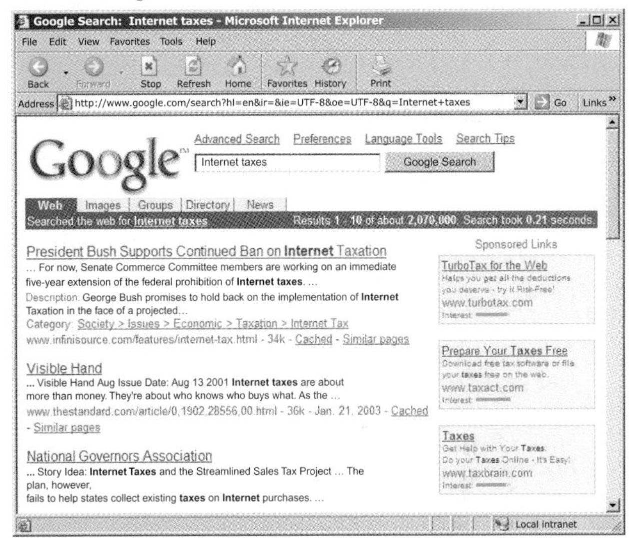

2. Second *Google* results

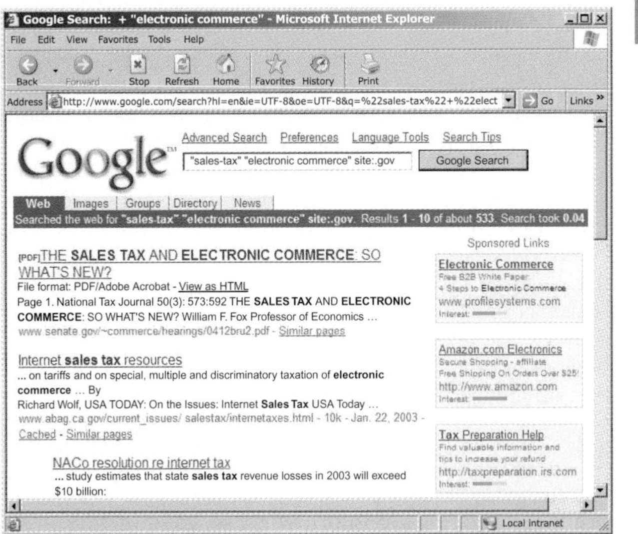

32e

of her keywords, specifying government-sponsored sites. Narrowed in this way, Alder's search produced a more manageable 533 hits—still a large number but including many potential sources on the first few screens.

32f ▪ Online communication

Several online sources can put you directly in touch with experts and others whose ideas and information may inform your research. Because these sources, like Web sites, are unfiltered, you must always evaluate them carefully. (See p. 123.)

- *Electronic mail* allows you to communicate with others who may be interested in your subject, such as a teacher at your school or an expert in another state. (See pp. 205–07 for more on e-mail.)
- *Discussion lists* (or listservs) use e-mail to connect subscribers who are interested in a common subject. Many have a scholarly or technical purpose. For an index of discussion lists, see *http://tile.net/lists* or *http://www.topica.com*.
- *Newsgroups and Web forums* are more open than discussion lists. For an index to Web forums, see *http://www.delphiforums.com*. For a newsgroup index, see *http://groups.google.com*.
- *Real-time, or synchronous, communication* allows you to participate in "live" conversation. For more on this resource, see *http://www.du.org/cybercomp.html* or *http://www.internet101.org/chat.html*.

32g ▪ Government publications

Government publications provide a vast array of data, public records, and other historical and contemporary information. For US government publications, consult the *Monthly Catalog of US Government Publications,* available on computer. Many federal, state, and local government agencies post important publications—reports, legislation, press releases—on their own Web sites. You can find lists of sites for various federal agencies by using the keywords *United States federal government* with a search engine. In addition, several Web sites are useful resources, such as *http://www.fedstats.gov* (government statistics) and *http://infoplease.com/us.html* (links to information on federal, state, and local governments).

32h ▪ Your own sources

Academic writing will often require you to conduct primary research for information of your own. For instance, you may need to analyze a poem, conduct an experiment, survey a group of people, or interview an expert.

An interview can be especially helpful for a research project because it allows you to ask questions precisely geared to your subject. You can conduct an interview in person, over the telephone, or online using electronic mail or a form of synchronous communication (see the facing page). A personal interview is preferable if you can arrange it, because you can hear the person's tone and see his or her expressions and gestures.

Here are a few guidelines for interviews:

- *Prepare a list of open-ended questions to ask*—perhaps ten or twelve for a one-hour interview. Do some research for these questions to discover background on the issues and your subject's published views on the issues.
- *Give your subject time to consider your questions.* Don't rush into silences with more questions.
- *Take care in interpreting answers,* especially if you are online and thus can't depend on facial expressions, gestures, and tone of voice to convey the subject's attitudes.
- *Keep thorough notes.* Take notes during an in-person or telephone interview, or record the interview if your subject agrees. For online interviews, save the discussion in a file of its own.
- *Verify quotations.* Before you quote your subject in your paper, check with him or her to ensure that the quotations are accurate.

33 Evaluating and Synthesizing Sources

Research writing is much more than finding sources and reporting their contents. The challenge and interest come from *selecting* appropriate sources and *interacting* with them through **critical reading**. To read critically, you analyze a text, identifying its main ideas, evidence, bias, and other relevant elements; you evaluate its usefulness and quality; and you relate it to other texts and to your own ideas.

Visit *www.ablongman.com/littlebrown* for added help and exercises on evaluating and synthesizing sources.

33a • Evaluation of sources

For most projects you will seek the mix of sources described on pages 106–08. But not all the sources in your working bibliography will contribute to this mix. Some may prove irrelevant to your subject; others may prove unreliable.

In evaluating sources you need to consider how they come to you. The sources you find through the library, both print and online, have been previewed for you by their publishers and by the library's staff. They still require your critical reading, but you can have some confidence in the information they contain. With Internet sources, however, you can't assume similar previewing, so your critical reading must be especially rigorous.

Library sources

To evaluate sources you find through the library, look at dates, titles, summaries, introductions, headings, and author biographies. Try to answer the following questions about each source.

Evaluate relevance.

- *Does the source devote some attention to your subject?* Check whether the source focuses on your subject or covers it marginally, and compare the source's coverage to that in other sources.
- *Is the source appropriately specialized for your needs?* Check the source's treatment of a topic you know something about, to ensure that it is neither too superficial nor too technical.
- *Is the source up to date enough for your subject?* Check the publication date. If your subject is current, your sources should be, too.

Evaluate reliability.

- *Where does the source come from?* It matters whether you found the source through your library or directly on the Internet. (If the latter, see opposite.) Check whether a library source is popular or scholarly. Scholarly sources, such as refereed journals and university press books, are generally deeper and more reliable.
- *Is the author an expert in the field?* The authors of scholarly publications tend to be experts. To verify expertise, check an author's credentials in a biography (if the source includes one), in a biographical reference, or by a keyword search of the Web.

- *What is the author's bias?* Every author has a point of view that influences the selection and interpretation of evidence. How do the author's ideas relate to those in other sources? What areas does the author emphasize, ignore, or dismiss? When you're aware of sources' biases, you can attempt to balance them.
- *Is the source fair and reasonable?* Even a strongly biased work should present sound reasoning, adequate evidence, and a fair picture of opposing views—all in an objective, calm tone. The absence of any of these qualities should raise a warning flag.
- *Is the source well written?* A coherent organization and clear, error-free sentences indicate a careful author.

Internet sources

To a great extent, the same critical reading that serves you with library sources will help you evaluate Internet sources that you reach directly. But Web sites and discussion-group postings can range from scholarly works to corporate promotions, from government-sponsored data to the self-published rantings of crackpots. You'll need to tell the difference.

To evaluate a Web site, add the following questions to those opposite for library sources. You can adapt these questions as well to discussion-group postings.

- *What does the URL lead you to expect about the site?* Are those expectations fulfilled?
- *Who is the author or sponsor?* How credible is the person or group responsible for the site?
- *What is the purpose of the site?* What does the site's author or sponsor intend to achieve?
- *What does context tell you?* What do you already know about the site's subject that can inform your evaluation? What kinds of support or other information do the site's links provide?
- *What does presentation tell you?* Is the site's design well thought out and effective? Is the writing clear and error-free?
- *How worthwhile is the content?* Are the site's claims well supported by evidence? Is the evidence from reliable sources?

In the next few pages we'll ask these questions of the Web site titled *Global Warming Information Center,* which turned up in a search for opinions and data on global warming.

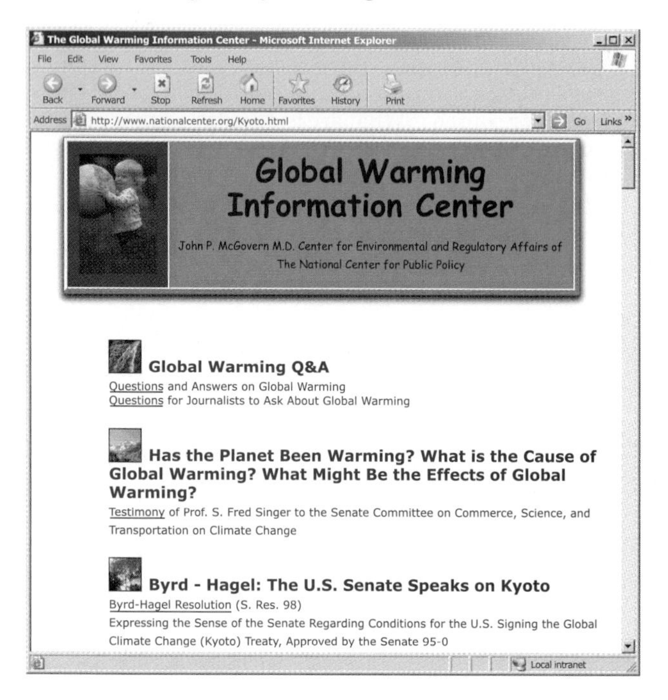

Check the URL.

In the screen shot, the URL (seen in the Address field) is *http://www.nationalcenter.org/Kyoto.html.* For purposes of evaluation, the most important part of any URL is the domain name—here, *nationalcenter.org*—which generally contains the name of the organization that sponsors the site (*nationalcenter*) and an abbreviation that describes the type of organization (*org*). *Org* designates a nonprofit organization. The other major abbreviations are *edu* (educational institution), *gov* (government body), *mil* (military), and *com* (commercial organization).

The domain abbreviation can inform your evaluation to some extent: a *com* site usually reflects the company's commercial purposes, an *edu* site usually supports and distributes scholarly pursuits, and an *org* site usually centers on the public interest. But the abbreviation should not unduly influence your evaluation. A *com* site may offer reliable data, an *edu* site may contain unfiltered student work, and an *org* site may promote a bigoted agenda.

Identify the author or sponsor.

A reputable site will list the author or group responsible for the site and will provide information or a link to that party. If none of this information is provided, you should not use the source. If you have only the author or group name, you may be able to discover more in a biographical dictionary or through a keyword search. You should also look for mentions of the author or group in your other sources.

The Web site *Global Warming Information Center* names its sponsor right up front: the John P. McGovern M.D. Center for Environmental and Regulatory Affairs, which is part of the National Center for Public Policy Research (the *nationalcenter* of the site's domain name). Discovering information about the McGovern Center requires scrolling to the end of the long home page and clicking on a link.

Gauge purpose.

A Web site's purpose determines what ideas and information it offers. Inferring that purpose tells you how to interpret what you see on the site. If a site is intended to sell a product or an opinion, it will likely emphasize favorable ideas and information while ignoring or even distorting what is unfavorable. In contrast, if a site is intended to build knowledge—for instance, a scholarly project or journal—it will likely acknowledge diverse views and evidence.

Determining the purpose of a site often requires looking beneath the surface of words and images and beyond the first page. The first screen of the *Global Warming* page—the title, the green banner, the photo of a child carrying a globe, the links to objective-sounding reports—suggest an environmentalist purpose of informing readers about the theory and consequences of rising earth temperatures caused by pollution. The site's purpose is actually different, though. A link to the McGovern Center's home page explains that the center's purpose is "to counter misinformation being spread to the public and policymakers by the environmental left" and "to . . . arm conservatives with tools for the environmental policy debate it had been lacking."

Consider context.

Look outside the site itself. What do you already know about the site's subject and the prevailing views of it? Do the site's links support the site's credibility? Are they relevant to the site and reliable in themselves?

Examining the *Global Warming* site, you might register the antiregulatory bias but also recognize that this view is a significant one in the debates over global warming. The question then is the reliability of its information: does it come from trustworthy, less-biased sources?

All the site's links lead to the McGovern Center's publications or to the parent organization, the National Center for Public Policy Research, so the question of reliability remains open.

Look at presentation.

Considering both the look of a site and the way it's written can illuminate its intentions and reliability. Does the design reflect the apparent purpose of the site, or does it undercut or conceal that purpose in some way? Is the text clearly written, or is it difficult to understand?

As noted earlier, the *Global Warming* site casts a "green" image that turns out not to coincide with its purpose. Otherwise, the clean design and straightforward, readable text indicate that the sponsor takes its purpose seriously and has thought out its presentation.

Analyze content.

With information about a site's author, purpose, and context, you're in a position to evaluate its content. Are the ideas and information slanted and, if so, in what direction? Are the views and data authoritative, or do you need to balance them—or even reject them? These questions require close reading of the text and its sources.

The *Global Warming* site links to a wealth of information on the issue it addresses, including many detailed reports. These documents offer statistics and quotations to support a skeptical view of global warming, with footnotes documenting sources. The footnotes are the crux of the site's reliability: the listed sources should be scholarly and should explain the methods of gathering and interpreting the cited data. Instead, however, they are newspaper and magazine articles and other nonscholary reports from the National Center and groups with similar aims.

The public controversy on global warming reflects disagreement among scholars over what causes global warming and whether it's even a potential problem. But the *Global Warming* site does not offer or refer to the scholarly research, so its claims and evidence must be viewed suspiciously and probably rejected for use in a research paper. A usable source need not be less biased, but it must be more substantial.

33b ▪ Synthesis of sources

Evaluating sources moves you into the most significant part of research writing: forging relationships for your own purpose. This **synthesis** is an essential step in reading sources critically and in creating new knowledge.

Respond to sources.

Write down what your sources make you think. Do you agree or disagree with the author? Do you find his or her views narrow, or do they open up new approaches for you? Is there anything in the source that you need to research further before you can understand it? Does the source prompt questions that you should keep in mind while reading other sources?

Connect sources.

When you notice a link between sources, jot it down. Do two sources differ in their theories or their interpretations of facts? Does one source illuminate another—perhaps commenting or clarifying or supplying additional data? Do two or more sources report studies that support a theory you've read about or an idea of your own?

Heed your own insights.

Apart from ideas prompted by your sources, you are sure to come up with independent thoughts: a conviction, a point of confusion that suddenly becomes clear, a question you haven't seen anyone else ask. These insights may occur at unexpected times, so it's good practice to keep a notebook handy to record them.

Use sources to support your own ideas.

As your research proceeds, the responses, connections, and insights you form through synthesis will lead you to answer your starting research question with a statement of your thesis (see p. 105). They will also lead you to the main ideas supporting your thesis—conclusions you have drawn from your synthesis of sources, forming the main divisions of your paper. When drafting the paper, make sure each paragraph focuses on an idea of your own, with supporting source material sandwiched between your explanations of it. Then your paper will synthesize others' work into something wholly your own.

34 Integrating Sources into Your Text

The evidence of others' information and opinions should back up, not dominate, your own ideas. To synthesize

Visit *www.ablongman.com/littlebrown* for added help and exercises on integrating sources into your text.

sources, you must decide whether to summarize, paraphrase, or quote them (below); you must introduce and interpret the borrowed material (p. 130); and you must clearly mark the boundaries of borrowed material (p. 135).

Note Integrating sources into your text may involve several conventions discussed elsewhere:

- For the punctuation of signal phrases such as *he insists,* see page 73.
- For guidelines on when to run quotations into your text and when to display them separately from your text, see pages 160 (MLA style), 176 (APA style), and 189 (Chicago style).
- For the use of the ellipsis mark (. . .) to indicate omissions from quotations, see pages 85–87.
- For the use of brackets ([ ]) to indicate changes in or additions to quotations, see pages 87 and 131.

34a ▪ Summary, paraphrase, and direct quotation

As you take notes from sources or work source material into your draft, you can summarize, paraphrase, quote, or combine methods. The choice should depend on why you are using a source.

Note *Summaries, paraphrases, and quotations all require source citations. A summary or paraphrase without a source citation or a quotation without quotation marks is plagiarism.* (See pp. 136–41 for more on plagiarism.)

Summary

When you **summarize,** you condense an extended idea or argument into a sentence or more in your own words. Summary is most useful when you want to record the gist of an author's idea without the background or supporting evidence. Here, for example, is a passage summarized in a sentence.

Original quotation

The following examples highlight the breadth of the digital divide today:

- Those with a college degree are more than *eight times* as likely to have a computer at home, and nearly *sixteen times* as likely to have home Internet access, as those with an elementary school education.
- A high-income household in an urban area is more than *twenty times* as likely as a rural, low-income household to have Internet access.

- A child in a low-income white family is *three times* as likely to have Internet access as a child in a comparable black family, and *four times* as likely to have access as children in a comparable Hispanic household.

—US Department of Commerce, *Falling Through the Net: Toward Digital Inclusion,* p. 7

Summary

US residents who are urban, white, college educated, and affluent are *much* more likely to be connected to the Internet than those who are rural, black or Hispanic, not educated past elementary school, and poor.

Paraphrase

When you **paraphrase**, you follow much more closely the author's original presentation, but you still restate it in your own words. Paraphrase is most useful when you want to reconstruct an author's line of reasoning but don't feel the original words merit direct quotation. Here is a paraphrase of the first two bulleted points in the quotation opposite from the Department of Commerce.

Paraphrase

Likelihood of being connected to the Internet among US groups:

Home connection, elementary education vs. college education: 1/16 as likely.

Any access, rural setting and low-income household vs. urban setting and affluent household: 1/20 as likely.

Follow these guidelines when paraphrasing:

- Read the material until you understand it.
- Restate the main ideas in your own words and sentence structures. Use phrases if complete sentences seem cumbersome.
- Be careful not to distort the author's meaning.

For examples of poor and revised paraphrases, see pages 139–40.

ESL If English is your second language, you may have difficulty paraphrasing the ideas in sources because synonyms don't occur to you or you don't see how to restructure sentences. Before attempting a paraphrase, read the original passage several times. Then, instead of "translating" line by line, try to state the gist of the passage without looking at it. Check your effort against the original to be sure you have captured the source author's meaning and emphasis without using his or her words and sentence structures. If you need a synonym for a word, look it up in a dictionary.

Direct quotation

If your purpose is to analyze a particular work, such as a short story or historical document, then you will use many direct quotations from the work. But otherwise you should quote from sources only in the following circumstances:

- *The author's original satisfies one of these requirements:*

  The language is unusually vivid, bold, or inventive.
  The quotation cannot be paraphrased without distortion or loss of meaning.
  The words themselves are at issue in your interpretation.
  The quotation represents and emphasizes the view of an important expert.
  The quotation is a graph, diagram, or table.

- *The quotation is as short as possible:*

  It includes only material relevant to your point.
  It is edited to eliminate examples and other unneeded material.

When taking a quotation from a source, copy the material *carefully.* Take down the author's exact wording, spelling, capitalization, and punctuation. Proofread every direct quotation *at least twice,* and be sure you have supplied big quotation marks so that later you won't confuse the direct quotation with a paraphrase or summary.

34b ▪ Introduction and interpretation of source material

Note Most examples in the following pages use the documentation style of the Modern Language Association (MLA) and also present-tense° verbs that are typical of much writing in the humanities. For specific variations in the academic disciplines, see pages 134–35.

Introduction

Work all quotations, paraphrases, and summaries smoothly into your own sentences, adding words as necessary to mesh structures.

Awkward One editor disagrees with this view and "a good reporter does not fail to separate opinions from facts" (Lyman 52).

Revised One editor disagrees with this view, maintaining that "a good reporter does not fail to separate opinions from facts" (Lyman 52).

To mesh your own and your source's words, you may sometimes need to make a substitution or addition to the quotation, signaling your change with brackets:

Words added

"The tabloids [of England] are a journalistic case study in bad reporting," claims Lyman (52).

Verb form changed

A bad reporter, Lyman implies, is one who "[fails] to separate opinions from facts" (52). [The bracketed verb replaces *fail* in the original.]

Capitalization changed

"[T]o separate opinions from facts" is a goal of good reporting (Lyman 52). [In the original, *to* is not capitalized.]

Noun supplied for pronoun

The reliability of a news organization "depends on [reporters'] trustworthiness," says Lyman (52). [The bracketed noun replaces *their* in the original.]

Interpretation

Even when it does not conflict with your own sentence structure, source material will be ineffective if you merely dump it in readers' laps without explaining how you intend it to be understood. In the following passage, we must figure out for ourselves that the writer's sentence and the quotation state opposite points of view.

Dumped Many news editors and reporters maintain that it is impossible to keep personal opinions from influencing the selection and presentation of facts. "True, news reporters, like everyone else, form impressions of what they see and hear. However, a good reporter does not fail to separate opinions from facts" (Lyman 52).

Revised Many news editors and reporters maintain that it is impossible to keep personal opinions from influencing the selection and presentation of facts. Yet not all authorities agree with this view. One editor grants that "news reporters, like everyone else, form impressions of what they see and hear." But, he insists, "a good reporter does not fail to separate opinions from facts" (Lyman 52).

Signal phrases

In the preceding revised passage, the words *One editor grants* and *he insists* are **signal phrases**: they tell readers who the source is and what to expect in the quotations. Signal phrases usually contain (1) the source author's name (or a substitute for it, such as *One editor* and *he*) and (2) a verb that indicates the source author's attitude or approach to what he or she says, as *grants* implies concession and *insists* implies argument.

Below are some verbs to use in signal phrases. For the appropriate tense° of such verbs (present,° as here, or past° or present perfect°) see pages 134–35.

Author is neutral	Author infers or suggests	Author argues	Author is uneasy or disparaging
comments	analyzes	claims	belittles
describes	asks	contends	bemoans
explains	assesses	defends	complains
illustrates	believes	disagrees	condemns
mentions	concludes	holds	deplores
notes	considers	insists	deprecates
observes	finds	maintains	derides
points out	predicts	**Author agrees**	laments
records	proposes	accepts	warns
relates	reveals	admits	
reports	shows	agrees	
says	speculates	concedes	
sees	suggests	concurs	
thinks	supposes	grants	
writes			

Note that some signal verbs, such as *describes* and *assesses,* cannot be followed by *that.*

Vary your signal phrases to suit your interpretation of source material and also to keep readers' interest. A signal phrase may precede, interrupt, or follow the borrowed material:

Signal phrase precedes

Lyman insists that "a good reporter does not fail to separate opinions from facts" (52).

Signal phrase interrupts

"However," Lyman insists, "a good reporter does not fail to separate opinions from facts" (52).

Signal phrase follows

"[A] good reporter does not fail to separate opinions from facts," Lyman insists (52).

Background information

You can add information to source material to integrate it into your text and inform readers why you are using it. Often, you may want to provide the author's name in the text:

Author named

Harold Lyman grants that "news reporters, like everyone else, form impressions of what they see and hear." But, Lyman insists, "a good reporter does not fail to separate opinions from facts" (52).

If the source title contributes information about the author or the context of the borrowed material, you can provide the title in the text:

Title given

Harold Lyman, in his book *The Conscience of the Journalist*, grants that "news reporters, like everyone else, form impressions of what they see and hear." But, Lyman insists, "a good reporter does not fail to separate opinions from facts" (52).

Finally, if the source author's background and experience reinforce or clarify the borrowed material, you can provide these credentials in the text:

Credentials given

Harold Lyman, a newspaper editor for more than forty years, grants that "news reporters, like everyone else, form impressions of what they see and hear." But, Lyman insists, "a good reporter does not fail to separate opinions from facts" (52).

You need not name the author, title, or credentials in your text when you are simply establishing facts or weaving together facts and opinions from varied sources to support a larger point. In the following passage, the information is more important than the source, so the author's name is confined to a parenthetical acknowledgment:

> To end the abuses of the British, many colonists were urging three actions: forming a united front, seceding from Britain, and taking control of their own international trade and diplomacy (Wills 325–36).

Discipline styles for interpreting sources

The preceding guidelines for interpreting source material apply generally across academic disciplines, but there are differences in verb tenses and documentation style.

English and some other humanities

Writers in English, foreign languages, and related disciplines use MLA style for documenting sources (see Chapter 36) and generally use the present tense° of verbs in signal phrases. (See the list of signal-phrase verbs on p. 132.) In discussing sources other than works of literature, the present perfect tense° is also sometimes appropriate:

> Lyman insists . . . [present].
> Lyman has insisted . . . [present perfect].

In discussing works of literature, use only the present tense to describe both the work of the author and the action in the work:

> Kate Chopin builds irony into every turn of "The Story of an Hour." For example, Mrs. Mallard, the central character, finds joy in the death of her husband, whom she loves, because she anticipates "the long procession of years that would belong to her absolutely" (23).

Avoid shifting tenses in writing about literature. You can, for instance, shorten quotations to avoid their past-tense verbs:

Shift — Her freedom elevates her, so that "she carried herself unwittingly like a goddess of victory" (24).

No shift — Her freedom elevates her, so that she walks "unwittingly like a goddess of victory" (24).

See page 160 for MLA format with poetry quotations and long prose quotations.

History and other humanities

Writers in history, art history, philosophy, and related disciplines generally use the present tense or present perfect tense of verbs in signal phrases. (See the list of possible verbs on page 132.)

> Lincoln persisted, as Haworth has noted, in "feeling that events controlled him."[3]

> What Miller calls Lincoln's "severe self-doubt"[6] undermined his effectiveness on at least two occasions.

The raised numbers after the quotations are part of the Chicago documentation style, used in history and other disciplines and discussed in Chapter 38.

See page 189 for the Chicago format with long prose quotations.

Social and natural sciences

Writers in the sciences generally use a verb's present tense just for reporting the results of a study (*The data suggest* . . .). Otherwise, they use a verb's past tense° or present perfect tense in a signal phrase, as when introducing an explanation, interpretation, or other commentary. (Thus, when you are writing for the sciences, generally convert the list of signal-phrase verbs on p. 132 from the present to the past or present perfect tense.)

> In an exhaustive survey of the literature published between 1990 and 2000, Walker (2001) found "no proof, merely a weak correlation, linking place of residence and rate of illness" (p. 121).
>
> Lin (1999) has suggested that preschooling may significantly affect children's academic performance through high school (pp. 22–23).

These passages conform to the documentation style of the American Psychological Association (APA), discussed in Chapter 37. APA style or the similar CSE style (Chapter 39) is used in sociology, education, nursing, biology, and many other sciences.

For APA format with long prose quotations, see page 176.

34c ▪ Clear boundaries for source material

Position source citations in your text to accomplish two goals: (1) make it clear exactly where your borrowing of source material begins and ends; (2) keep the citations as unobtrusive as possible. You can accomplish both goals by placing an in-text citation at the end of the sentence element containing the borrowed material. This sentence element may be a phrase or a clause, and it may begin, interrupt, or conclude the sentence.

> The inflation rate might climb as high as 30 percent (Kim 164), an increase that could threaten the small nation's stability.
>
> The inflation rate, which might climb as high as 30 percent (Kim 164), could threaten the small nation's stability.
>
> The small nation's stability could be threatened by its inflation rate, which, one source predicts, might climb as high as 30 percent (Kim 164).

In the last example the addition of *one source predicts* clarifies that Kim is responsible only for the inflation-rate prediction, not for the statement about stability.

When your paraphrase or summary of a source runs longer than a sentence, clarify the boundaries by using the author's name in the first sentence and placing the parenthetical citation at the end of the last sentence.

Juliette Kim studied the effects of acutely high inflation in several South American and African countries since World War II. She discovered that a major change in government accompanied or followed the inflationary period in 56 percent of cases (22-23).

35 Avoiding Plagiarism and Documenting Sources

Plagiarism (from a Latin word for "kidnapper") is the presentation of someone else's ideas or words as your own. Whether deliberate or accidental, plagiarism is a serious and often punishable offense.

- *Deliberate plagiarism* includes downloading a sentence from a source and passing it off as your own, summarizing someone else's ideas without acknowledging your debt, or buying a term paper and handing it in as your own.
- *Accidental plagiarism* includes forgetting to place quotation marks around another writer's words, omitting a source citation because you are not aware of the need for it, or carelessly copying a source when you mean to paraphrase.

ESL More than in many other cultures, teachers in the United States value students' original thinking and writing. In some other cultures, for instance, students may be encouraged to copy the words of scholars without acknowledgment, in order to demonstrate their mastery of or respect for the scholars' work. But in the United States any use of another's words or ideas without a source citation is plagiarism and is unacceptable. When in doubt about the guidelines in this section, ask your instructor for advice.

Visit *www.ablongman.com/littlebrown* for added help and an exercise on avoiding plagiarism.

CHECKLIST FOR AVOIDING PLAGIARISM

Type of source

- ☑ Are you using (1) your own independent material, (2) common knowledge, or (3) someone else's independent material? *You must acknowledge someone else's material.*

Quotations

- ☑ Do all quotations exactly match their sources? Check them.
- ☑ Have you inserted quotation marks around quotations that are run into your text?
- ☑ Have you shown omissions with ellipsis marks and additions with brackets?
- ☑ Does every quotation have a source citation?

Paraphrases and summaries

- ☑ Have you used your own words and sentence structures for every paraphrase and summary? If not, use quotation marks around the original author's words.
- ☑ Does every paraphrase and summary have a source citation?

The Web

- ☑ Have you obtained any necessary permission to use someone else's material on your Web site?

Source citations

- ☑ Have you acknowledged every use of someone else's material in the place where you use it?
- ☑ Does your list of works cited include all the sources you have used?

35a ▪ Plagiarism and the Internet

The Internet has made it easier to plagiarize than ever before, but it has also made plagiarism easier to catch.

Even honest students risk accidental plagiarism by downloading sources and importing portions into their drafts. Dishonest students may take advantage of downloading to steal others' work. They may also use the term-paper businesses on the Web, which offer both ready-made research and complete papers, usually for a fee. *Paying for research or a paper does not make it the buyer's work.* Anyone who submits someone else's work as his or her own is a plagiarist.

Students who plagiarize from the Internet both deprive themselves of an education in honest research and expose themselves to detection. Teachers can use search engines to locate specific phrases or sentences anywhere on the Web, including among scholarly publications, all kinds of Web sites, and term-paper collections. They can search the term-paper sites as easily as students can, looking for similarities with papers they have received. Increasingly, teachers can use plagiarism-detection programs that compare students' work with other work anywhere on the Internet, seeking matches as short as a few words.

Some instructors suggest that their students use plagiarism-detection programs to verify that their own work does not include accidental plagiarism, at least not from the Internet.

35b ▪ What *not* to acknowledge

Your independent material

You are not required to acknowledge your own observations, thoughts, compilations of facts, or experimental results, expressed in your own words and format.

Common knowledge

You need not acknowledge common knowledge: the standard information of a field of study as well as folk literature and commonsense observations.

If you do not know a subject well enough to determine whether a piece of information is common knowledge, make a record of the source. As you read more about the subject, the information may come up repeatedly without acknowledgment, in which case it is probably common knowledge. But if you are still in doubt when you finish your research, always acknowledge the source.

35c ▪ What *must* be acknowledged

You must always acknowledge other people's independent material—that is, any facts or ideas that are not common knowledge or your own. The source may be anything, including a book, an article, a movie, an interview, a microfilmed document, a Web page, a computer program, a newsgroup posting, or an opinion expressed on the radio. You must acknowledge summaries or paraphrases of ideas or facts as well as quotations of the language and format in which ideas or facts appear: wording, sentence structures, arrangement, and special graphics (such as a diagram). You must acknowledge another's material no matter how you use it, how much of it you use, or how often you use it.

Note See pages 127–31 on integrating quotations into your own text without plagiarism. And see pages 141–42 on acknowledging sources.

Copied language: Quotation marks and a source citation

The following example baldly plagiarizes the original quotation from Jessica Mitford's *Kind and Usual Punishment*, page 9. Without quotation marks or a source citation, the example matches Mitford's wording (underlined) and closely parallels her sentence structure:

Original quotation — "The character and mentality of the keepers may be of more importance in understanding prisons than the character and mentality of the kept."

Plagiarism — But the character of prison officials (the keepers) is more important in understanding prisons than the character of prisoners (the kept).

To avoid plagiarism, the writer has two options: (1) paraphrase and cite the source (see the examples on the next page) or (2) use Mitford's actual words *in quotation marks* and *with a source citation* (here, in MLA style):

Revision (quotation) — According to one critic of the penal system, "The character and mentality of the keepers may be of more importance in understanding prisons than the character and mentality of the kept" (Mitford 9).

Even with a source citation and with a different sentence structure, the next example is still plagiarism because it uses some of Mitford's words (underlined) without quotation marks:

35c

Plagiarism — According to one critic of the penal system, the psychology of the kept may say less about prisons than the psychology of the keepers (Mitford 9).

Revision (quotation) — According to one critic of the penal system, the psychology of "the kept" may say less about prisons than the psychology of "the keepers" (Mitford 9).

Paraphrase or summary: Your own words and a source citation

The example below changes Mitford's sentence structure, but it still uses her words (underlined) without quotation marks and without a source citation:

Plagiarism — In understanding prisons, we should know more about the character and mentality of the keepers than of the kept.

To avoid plagiarism, the writer can use quotation marks and cite the source (previous page) or *use his or her own words* and still *cite the source* (because the idea is Mitford's, not the writer's):

Revision (paraphrase) Mitford holds that we may be able to learn more about prisons from the psychology of the prison officials than from that of the prisoners (9).

Revision (paraphrase) We may understand prisons better if we focus on the personalities and attitudes of the prison workers rather than those of the inmates (Mitford 9).

In the next example, the writer cites Mitford and does not use her words but still plagiarizes her sentence structure:

Plagiarism One critic of the penal system maintains that the psychology of prison officials may be more informative about prisons than the psychology of prisoners (Mitford 9).

Revision (paraphrase) One critic of the penal system maintains that we may be able to learn less from the psychology of prisoners than from the psychology of prison officials (Mitford 9).

35d ▪ Online sources

35d

You should acknowledge online sources when you would any other source: whenever you use someone else's independent material in any form. But online sources may present additional challenges as well:

- *Record complete source information* as noted on pages 109–11 each time you consult the source. Online sources may change from one day to the next or even be removed entirely. Without the source information, you *may not* use the source.
- *Acknowledge linked sites.* The fact that one person has used a second person's work does not release you from the responsibility to acknowledge the second work.
- *Seek the author's permission* to use any ideas, information, or wording you obtain from e-mail correspondence or a discussion group.

If you want to use material in something you publish online, such as your own Web site, seek permission from the copyright holder in addition to citing the source. Generally, you can find information about a site's copyright on the home page or at the bottoms of other pages: look for a notice using the symbol ©. Most worthwhile sites also provide information for contacting the author or sponsor. (See p. 111 for an illustration.) If you don't find a copyright notice, you *cannot* assume that the work is unprotected by copyright. Only if the site explicitly says it is not copyrighted or is available for free use can you use it online without permission.

35e ▪ Documentation of sources

Every time you borrow the words, facts, or ideas of others, you must **document** the source—that is, supply a reference (or document) telling readers that you borrowed the material and where you borrowed it from. (For when to document sources, see pp. 138–40.)

Editors and teachers in most academic disciplines require special documentation formats (or styles) in their scholarly journals and in students' papers. All the styles use a citation in the text that serves two purposes: it signals that material is borrowed, and it refers readers to detailed information about the source so that they can locate both the source and the place in the source where the borrowed material appears. The detailed source information appears either in footnotes or at the end of the paper.

Aside from these essential similarities, the disciplines' documentation styles differ markedly in citation form, arrangement of source information, and other particulars. Each discipline's style reflects the needs of its practitioners for certain kinds of information presented in certain ways. For instance, the currency of a source is important in the social and natural sciences, where studies build on and correct each other; thus in-text citations in these disciplines usually include a source's date of publication. In the humanities, however, currency is less important, so in-text citations do not include date of publication.

The documentation formats of the disciplines are described in style guides, including those in the following list. This book presents the styles of the guides that are marked *.

Humanities

**The Chicago Manual of Style.* 15th ed. 2003. (See pp. 178–91.)

*Gibaldi, Joseph. *MLA Handbook for Writers of Research Papers.* 6th ed. 2003. (See pp. 142–62.)

Social sciences

American Anthropological Association. "AAA Style Guide." 2003. *http://www.aaanet.org/pubs/style_guide.htm.*

American Political Science Association. *Style Manual for Political Science.* 2001.

*American Psychological Association. *Publication Manual of the American Psychological Association.* 5th ed. 2001. (See pp. 163–78.)

American Sociological Association. *ASA Style Guide.* 2nd ed. 1997.

A Uniform System of Citation (law). 16th ed. 1996.

Sciences and mathematics

American Chemical Society. *ACS Style Guide: A Manual for Authors and Editors.* 2nd ed. 1997.

American Institute of Physics. *Style Manual for Guidance in the Preparation of Papers.* 4th ed. 1997.

American Medical Association Manual of Style. 9th ed. 1998.

Bates, Robert L., Rex Buchanan, and Marla Adkins-Heljeson, eds. *Geowriting: A Guide to Writing, Editing, and Printing in Earth Science.* 5th ed. 1995.

*Council of Biology Editors. *Scientific Style and Format: The CBE Manual for Authors, Editors, and Publishers.* 6th ed. 1994. (CBE is now called the Council of Science Editors.) (See pp. 191–98.)

Ask your instructor or supervisor which style you should use. If no style is required, use the guide from the preceding list that's most appropriate for the discipline in which you're writing. Do follow one system for citing sources—and one system only—so that you provide all the necessary information in a consistent format.

MLA 36

Note Various computer programs can help you format your source citations in the style of your choice. The programs prompt you for needed information (author's name, book title, and so on) and then format the information as required by the style. The programs remove some tedium from documenting sources, but they can't substitute for your own care and attention in giving your sources accurate and complete acknowledgment.

36 MLA Documentation and Format

Widely used in English, foreign languages, and other humanities, the MLA documentation style originates with

Visit *www.ablongman.com/littlebrown* for added help and an exercise on MLA documentation style.

MLA PARENTHETICAL TEXT CITATIONS

the Modern Language Association. It appears in *MLA Handbook for Writers of Research Papers*, 6th ed. (2003). The MLA's Web site, at *http://www.mla.org*, offers occasional updates and answers to frequently asked questions about MLA style.

In MLA style, brief parenthetical citations in the text (below) direct readers to a list of works cited at the end of the text (p. 147). This chapter also details the MLA format for papers (p. 159).

36a ▪ MLA parenthetical text citations

Citation formats

The in-text citations of sources must include just enough information for the reader to locate (1) the appropriate source in your list of works cited and (2) the place in the source where the borrowed material appears. Usually, you can meet both these requirements by providing the author's last name and the page(s) in the source on which the borrowed material appears.

Note Examples 1 and 2 on the next page show the two ways of writing in-text citations. In examples 3–13 any citation with the author or title named in the text could be written instead with the author or title named in parentheses, and vice versa. The choice depends on how you want to introduce and interpret your sources (see pp. 130–35).

1. Author named in your text

One researcher, Carol Gilligan, concludes that "women impose a distinctive construction on moral problems, seeing moral dilemmas in terms of conflicting responsibilities" (105).

2. Author not named in your text

One researcher concludes that "women impose a distinctive construction on moral problems, seeing moral dilemmas in terms of conflicting responsibilities" (Gilligan 105).

3. A work with two or three authors

As Frieden and Sagalyn observe, "The poor and the minorities were the leading victims of highway and renewal programs" (29).

One text discusses the "ethical dilemmas in public relations practice" (Wilcox, Ault, and Agee 125).

4. A work with more than three authors

Give all authors' names, or give only the first author's name followed by "et al." (the abbreviation for the Latin meaning "and others"). Do the same in your list of works cited.

Lopez, Salt, Ming, and Reisen observe that it took the combined forces of the Americans, Europeans, and Japanese to break the rebel siege of Beijing in 1900 (362).

5. A work with numbered paragraphs or screens instead of pages

According to Palfrey, twins raised apart report similar feelings (pars. 6-7).

6. An entire work or a work with no page or other numbers

Almost 20 percent of commercial banks have been audited for the practice (Friis).

7. A multivolume work

If you name more than one volume in your list of works cited, give the volume you're acknowledging in your text citation (here, volume 5).

After issuing the Emancipation Proclamation, Lincoln said, "What I did, I did after very full deliberations, and under a very heavy and solemn sense of responsibility" (5: 438).

If you name only one volume in your list of works cited, give only the page number in your text citation.

8. A work by an author of two or more cited works

Give the title of the particular work you're citing, shortening it in a parenthetical citation unless it's already brief. In the following example, *Arts* is short for the full title, *The Arts and Human Development*.

At about age seven, most children begin to use appropriate gestures to reinforce their stories (Gardner, Arts 144-45).

9. An unsigned work

Use the title (or a brief version of it) in place of an author's name. This citation has no page number because the entire source is only a page.

"The Right to Die" notes that a death-row inmate may demand his own execution to achieve a fleeting notoriety.

10. A government publication or a work with a corporate author

One study found that teachers of secondary social studies are receiving more training in methods but less in the discipline itself (Lorenz Research 64).

11. A source referred to by another source

George Davino maintains that "even small children have vivid ideas about nuclear energy" (qtd. in Boyd 22).

12. A literary work

For novels that may be available in many editions, cite part, chapter, or other numbers as well as page number.

Toward the end of James's novel, Maggie suddenly feels "the intimate, the immediate, the familiar, as she hadn't had them for so long" (535; pt. 6, ch. 41).

For verse plays and poems that are divided into parts, omit a page number. Instead, as in the next example, cite the part (act 3, scene 4) and line number(s) (147).

Later in King Lear Shakespeare has the disguised Edgar say, "The prince of darkness is a gentleman" (3.4.147).

13. The Bible

Abbreviate any book title longer than four letters, and then give chapter and verse(s).

According to the Bible, at Babel God "did . . . confound the language of all the earth" (Gen. 11.9).

14. An electronic source

Many electronic sources can be cited just as printed sources are. For a source with no named author, see model 9. For a source that uses paragraph or some other numbers instead of page numbers, see model 5. For a source with no numbering, just give the author's name, either in parentheses, as in model 6, or in the text:

Michael Tourville, for one, believes that Europe's single currency will strengthen the continent's large and technologically advanced companies.

15. Two or more works in the same citation

Two recent articles point out that a computer badly used can be less efficient than no computer at all (Gough and Hall 201; Richards 162).

Footnotes or endnotes in special circumstances

Footnotes or endnotes may supplement parenthetical citations when you cite several sources at once, when you comment on a source, or when you provide information that does not fit easily in the text. Signal a footnote or endnote in your text with a numeral raised above the appropriate line. Then write a note with the same numeral.

Text

At least five studies have confirmed these results.[1]

Note

[1] Abbott and Winger 266-68; Casner 27; Hoyenga 78-79; Marino 36; Tripp, Tripp, and Walk 179-83.

In a note the raised numeral is indented one-half inch or five spaces and is followed by a space. If the note appears as a footnote, place it at the bottom of the page on which the citation appears, set it off from the text with quadruple spacing, and single-space the note itself. If the note appears as an endnote, place it in numerical order with the other endnotes on a page between the text and the list of works cited; double-space all the endnotes.

36b ▪ MLA list of works cited

On a new page at the end of your paper, a list titled "Works Cited" includes all the sources you quoted, paraphrased, or summarized in your paper. The format of the list is described below and illustrated on page 162.

Arrangement Arrange your sources in alphabetical order by the last name of the author (the first author if there is more than one). If an author is not given in the source, alphabetize the source by the first main word of the title (excluding *A*, *An*, or *The*).

Spacing and indention Double-space all entries. Type the first line of each entry at the left margin, and indent all subsequent lines one-half inch or five spaces.

Authors List the author's name last-name first. If there are two or more authors, list all names after the first in normal order. Separate the names with commas.

Title Give full titles, capitalizing all important words (see p. 97). For periodical titles, omit any initial *A*, *An*, or *The*. Unless your instructor specifies italics, underline the titles of books and periodicals; place titles of periodical articles in quotation marks.

Publication information Provide the publication information after the title. For books, give place of publication, publisher's name, and date. (Shorten publishers' names—for instance, *Little* for Little, Brown and *Harvard UP* for Harvard University Press.) For periodical articles, give the volume or issue number, date, and page numbers.

Punctuation Separate the main parts of an entry with periods followed by one space.

MLA 36b

Books

1. A book with one author

Gilligan, Carol. In a Different Voice: Psychological Theory and Women's Development. Cambridge: Harvard UP, 1982.

2. A book with two or three authors

Lifton, Robert Jay, and Greg Mitchell. Who Owns Death: Capital Punishment, the American Conscience, and the End of Executions. New York: Morrow, 2000.

MLA WORKS-CITED MODELS

Books

Periodicals

Electronic sources

Wilcox, Dennis L., Phillip H. Ault, and Warren K. Agee. Public Relations: Strategies and Tactics. 4th ed. New York: Harper, 1999.

3. A book with more than three authors

Give all authors' names, or give only the first author's name followed by "et al." (the abbreviation for the Latin meaning "and others"). Do the same in your in-text citations of the source.

Lopez, Geraldo, Judith P. Salt, Anne Ming, and Henry Reisen. China and the West. Boston: Little, 1995.

Lopez, Geraldo, et al. China and the West. Boston: Little, 1995.

4. Two or more works by the same author(s)

Gardner, Howard. The Arts and Human Development. New York: Wiley, 1973.

---. The Quest for Mind: Piaget, Lévi-Strauss, and the Structuralist Movement. New York: Knopf, 1973.

5. A book with an editor

Holland, Merlin, and Rupert Hart-Davis, eds. The Complete Letters of Oscar Wilde. New York: Holt, 2000.

6. A book with an author and an editor

Mumford, Lewis. The City in History. Ed. Donald L. Miller. New York: Pantheon, 1986.

7. A translation

Alighieri, Dante. The Inferno. Trans. John Ciardi. New York: NAL, 1971.

8. A book with a corporate author

Lorenz Research, Inc. Research in Social Studies Teaching. Baltimore: Arrow, 2004.

9. An anonymous book

The Dorling Kindersley World Reference Atlas. London: Dorling, 2005.

10. The Bible

The Bible. King James Version.

The New English Bible. London: Oxford UP and Cambridge UP, 1970.

11. A later edition

Bolinger, Dwight L. Aspects of Language. 2nd ed. New York: Harcourt, 1975.

12. A republished book

James, Henry. The Golden Bowl. 1904. London: Penguin, 1966.

13. A book with a title in its title

Eco, Umberto. Postscript to The Name of the Rose. Trans. William Weaver. New York: Harcourt, 1983.

14. A work in more than one volume

Using two or more volumes:

Lincoln, Abraham. The Collected Works of Abraham Lincoln. Ed. Roy P. Basler. 8 vols. New Brunswick: Rutgers UP, 1953.

Using only one volume:

Lincoln, Abraham. The Collected Works of Abraham Lincoln. Ed. Roy P. Basler. Vol. 5. New Brunswick: Rutgers UP, 1953. 8 vols.

15. A work in a series

Bergman, Ingmar. The Seventh Seal. Mod. Film Scripts Ser. 12. New York: Simon, 1968.

16. An anthology

Kennedy, X. J., and Dana Gioia, eds. Literature: An Introduction to Fiction, Poetry, and Drama. 9th ed. New York: Longman, 2005.

17. A selection from an anthology

Allende, Isabel. "The Judge's Wife." Trans. Margaret Sayers Peden. Literature: An Introduction to Fiction, Poetry, and Drama. Ed. X. J. Kennedy and Dana Gioia. 9th ed. New York: Longman, 2005. 479-84.

A reprinted scholarly article:

Reekmans, Tony. "Juvenal on Social Change." Ancient Society 2 (1971): 117-61. Rpt. in Private Life in Rome. Ed. Helen West. Los Angeles: Coronado, 1981. 124-69.

18. Two or more selections from the same anthology

Chopin, Kate. "The Storm." Kennedy and Gioia 552-53.

Kennedy, X. J., and Dana Gioia, eds. Literature: An Introduction to Fiction, Poetry, and Drama. 9th ed. New York: Longman, 2005.

O'Connor, Flannery. "Revelation." Kennedy and Gioia 443-58.

19. An introduction, preface, foreword, or afterword

Donaldson, Norman. Introduction. The Claverings. By Anthony Trollope. New York: Dover, 1977. vii-xv.

20. An article in a reference work

Mark, Herman F. "Polymers." The New Encyclopaedia Britannica: Macropaedia. 16th ed. 1997.

"Reckon." Merriam-Webster's Collegiate Dictionary. 11th ed. 2003.

Periodicals: Journals, magazines, newspapers

21. An article in a journal with continuous pagination throughout the annual volume

Lever, Janet. "Sex Differences in the Games Children Play." Social Problems 23 (1996): 478-87.

22. An article in a journal that pages issues separately or that numbers only issues, not volumes

Dacey, June. "Management Participation in Corporate Buy-Outs." Management Perspectives 7.4 (1998): 20-31.

23. An article in a monthly or bimonthly magazine

Tilin, Andrew. "Selling the Dream." Worth Sept. 2001: 94-100.

24. An article in a weekly or biweekly magazine

Talbot, Margaret. "The Bad Mother." New Yorker 5 Aug. 2004: 40-46.

25. An article in a daily newspaper

Kolata, Gina. "Kill All the Bacteria!" New York Times 7 Jan. 2005, natl. ed.: B1+.

26. An anonymous article

"The Right to Die." Time 11 Oct. 1996: 101.

27. An editorial or letter to the editor

"Dueling Power Centers." Editorial. New York Times 14 Jan. 2005, natl. ed.: A16.

Dowding, Michael. Letter. Economist 5-11 Jan. 2005: 4.

28. A review

Nelson, Cary. "Between Anonymity and Celebrity." Rev. of Anxious Intellects: Academic Professionals, Public Intellectuals, and Enlightenment Values, by John Michael. College English 64 (2002): 710-19.

29. An abstract of a dissertation

Steciw, Steven K. "Alterations to the Pessac Project of Le Corbusier." Diss. U of Michigan, 1986. DAI 46 (1986): 565C.

30. An abstract of an article

Lever, Janet. "Sex Differences in the Games Children Play." Social Problems 23 (1996): 478-87. Psychological Abstracts 63 (1996): item 1431.

Electronic sources

Electronic sources include those available on CD-ROM and those available online, either through your library's Web site or directly over the Internet. Online sources require two special pieces of information:

- Give the date when you consulted the source as well as the date when the source was posted online. The posting date comes first, with other publication information. Your access date falls just before the electronic address.
- Give the source's exact electronic address, or URL, enclosed in angle brackets (< >) at the end of the entry. If you must break an address from one line to the next, do so *only* after a slash, and do not hyphenate.

Note A URL does not always provide a usable route to a source, especially when it is impossibly long or is unique to a particular search or a particular library. See models 42–43 (online subscription services) for examples of what to do in such a case.

Try to locate all the information required in the following models, referring to pages 110–11 if you need help. However, if you search for and still cannot find some information, then give what you can find.

31. A source on a periodical CD-ROM database

Lewis, Peter H. "Many Updates Cause Profitable Confusion." New York Times 21 Jan. 2004 natl. ed.: D1+. New York Times Ondisc. CD-ROM. UMI-ProQuest. Mar. 2004.

The final date is the CD-ROM's publication date.

32. A source on a nonperiodical CD-ROM

Shelley, Mary Wollstonecraft. Frankenstein. Classic Library. CD-ROM. Alameda: Andromeda, 1999.

33. An online book

An entire book:

Austen, Jane. Emma. Ed. Ronald Blythe. Harmondsworth: Penguin, 1972. Oxford Text Archive. 1998. Oxford U. 15 Dec. 2004 <http://ota.ox.ac.uk/pub/ota/public/english/Austen/emma.1519>.

A part of a book:

Conrad, Joseph. "A Familiar Preface." Modern Essays. Ed. Christopher Morley. New York: Harcourt, 1921. Bartleby.com: Great Books Online. Ed. Steven van Leeuwan. Nov. 2000. 16 Feb. 2005 <http://www.bartleby.com/237/8.html>.

34. An article in an online journal

Palfrey, Andrew. "Choice of Mates in Identical Twins." Modern Psychology 4.1 (1996): 26-40. 25 Feb. 2005 <http://www.liasu.edu/modpsy/palfrey4(1).htm>.

35. An online abstract

Palfrey, Andrew. "Choice of Mates in Identical Twins." Modern Psychology 4.1 (1996): 26-40. Abstract. 25 Feb. 2005 <http://www.liasu.edu/modpsy/abstractpalfrey4(1).htm>.

36. An article in an online newspaper

Still, Lucia. "On the Battlefields of Business, Millions of Casualties." New York Times on the Web 3 Mar. 2002. 17 Apr. 2004 <http://www.nytimes.com/specials/downsize/03down1.html>.

37. An article in an online magazine

Lewis, Ricki. "The Return of Thalidomide." Scientist 22 Jan. 2001: 5. 24 Jan. 2005 <http://www.the-scientist.com/yr2001/jan/lewis_pl_010122.html>.

38. An online review

Detwiler, Donald S., and Chu Shao-Kang. Rev. of Important Documents of the Republic of China, ed. Tan Quon Chin. Journal of Military History 56.4 (1992): 669-84. 16 Sept. 2004 <http://www.jstor.org/fcgi-bin/jstor/viewitem.fcg/08993718/96p0008x>.

39. An entire site (scholarly project, professional site, personal site, etc.)

Scots Teaching and Research Network. Ed. John Corbett. 2 Feb. 2003. U of Glasgow. 5 Mar. 2004 <http://www.arts.gla.ac.uk/www/comet/starn.htm>.

Lederman, Leon. Topics in Modern Physics--Lederman. 10 Oct. 2004. 12 Dec. 2004 <http://www.ed.fnal.gov/samplers/hsphys/people/lederman.html>.

40. A short work from an online site

Barbour, John. "The Brus." Scots Teaching and Research Network. Ed. John Corbett. 2 Feb. 2003. U of Glasgow. 5 Mar. 2004 <http://www.arts.gla.ac.uk/www/comet/starn/poetry/brus/contents.htm>.

41. The home page for a course

Anderson, Daniel. Business Communication. Course home page. Jan.-June 2004. Dept. of English, U of North Carolina. 16 Feb. 2004 <http://sites.unc.edu/daniel/eng32/index.html>.

42. A work from an online service to which your library subscribes

Netchaeva, Irina. "E-Government and E-Democracy." International Journal for Communication Studies 64 (2002): 467-78. Academic Search Elite. EBSCOhost. Santa Clara U, Orradre Lib. 20 Dec. 2004 <http://www.epnet.com>.

Your library subscribes to a number of online services that provide access to the full text of articles in periodicals, reference works, and other sources. For a source you find via a subscription service, follow the publication information with the name of the database you used (underlined), the name of the service (not underlined), and the name of the subscribing institution and library. If the source URL is temporary (generated for each search) or unique to the subscribing library, give the URL of the service's home page, if known, so that readers can locate information about the service. If you can't find this URL, you may end the entry with the date of your access.

43. A work from an online service to which you subscribe

"China--Dragon Kings." The Encyclopedia Mythica. America Online. 6 Jan. 2004. Path: Research and Learn; Encyclopedia; More Encyclopedias; Encyclopedia Mythica.

If you find a source through America Online, MSN, or another personal online service, you may not see a usable URL or any URL for the source. In that case, provide the path you used to get to the source (example above) or the keyword (for instance, Keyword: Chinese dragon kings).

44. An online government publication

United States. Dept. of Commerce. National Telecommunications and Information Admin. Falling Through the Net: Toward Digital Inclusion. Oct. 2001. 1 Mar. 2004 <http://www.ntia.doc.gov/ntiahome/fttn00/contents00.html>.

45. An article in an online information database

Pull, Jack L. "Wu-ti." Encyclopaedia Britannica Online. 2004. Encyclopaedia Britannica. 23 Dec. 2004 <http://www.eb.com:80>.

46. An online graphic, video, or audio source

Hamilton, Calvin J. "Components of Comets." Diagram. Space Art. 2002. 20 Dec. 2004 <http://solarviews.com/eng/comet.htm>.

Stewart, Leslie J. 96 Ranch Rodeo and Barbecue. 1951. Library of Congress. 7 Jan. 2005 <http://memory.loc.gov/cgi-bin/query/ammem/ncr:@field(DocID+/@lit(nv034))>.

Reagan, Ronald W. State of the Union address. 20 Jan. 1982. Vincent Voice Library. Digital and Multimedia Center. U of Michigan. 6 May 2004 <http://www.lib.msu.edu/vincent/presidents/reagan.htm>.

47. Electronic mail

Millon, Michele. "Re: Grief Therapy." E-mail to the author. 4 May 2004.

48. A posting to an e-mail discussion list

Tourville, Michael. "European Currency Reform." Online posting. 6 Jan. 2004. International Finance Archive. 12 Jan. 2005 <http://www.weg.isu.edu/finance-dl/060104/46732>.

49. A posting to a newsgroup or Web forum

Cramer, Sherry. "Recent Investment Practices." Online posting. 26 Mar. 2004. 3 Apr. 2004 <news:biz.investment.current.2700>.

Franklin, Melanie. Online posting. 25 Jan. 2005. The Creative Process. 27 Jan. 2005 <http://forums/nytimes.com/webin/WebX?14@^182943@.eea3ea7>.

50. A synchronous communication

Bruckman, Amy. MediaMOO Symposium: Virtual Worlds for Business? 20 Jan. 2005. MediaMOO. 26 Feb. 2005 <http://www.cc.gatech.edu/Amy.Bruckman/MediaMOO/cscw-symposium-98.html>.

51. Computer software

Project Scheduler 9000. Ver. 5.1. Orlando: Scitor, 2005.

Other sources

52. A government publication

Hawaii. Dept. of Education. Kauai District Schools, Profile 2003-04. Honolulu: Hawaii Dept. of Education, 2005.

Stiller, Ann. Historic Preservation and Tax Incentives. US Dept. of the Interior. Washington: GPO, 1996.

United States. Cong. House. Committee on Ways and Means. Medicare Payment for Outpatient Physical and Occupational Therapy Services. 108th Cong., 1st sess. Washington: GPO, 2003.

53. A musical composition or work of art

Fauré, Gabriel. Sonata for Violin and Piano no. 1 in A Major, op. 15.

Sargent, John Singer. Venetian Doorway. Metropolitan Museum of Art, New York. Sargent Watercolors. By Donelson F. Hoopes. New York: Watson, 1976. 31.

54. A film, video recording, or DVD

Schindler's List. Dir. Steven Spielberg. Perf. Liam Neeson and Ben Kingsley. Universal, 1993.

George Balanchine, chor. Serenade. Perf. San Francisco Ballet. Dir. Hilary Bean. 1981. Videocassette. PBS Video, 1987.

55. A television or radio program

"I'm Sorry, I'm Lost." By Alan Ball. Dir. Jill Soloway. Six Feet Under. HBO. 2 Sept. 2004.

56. A performance

The English Only Restaurant. By Silvio Martinez Palau. Dir. Susana Tubert. Puerto Rican Traveling Theater, New York. 27 July 2004.

Eddins, William, cond. Chicago Symphony Orch. Symphony Center, Chicago. 22 Jan. 2005.

57. A sound recording

Springsteen, Bruce. "Empty Sky." The Rising. Columbia, 2002.

Brahms, Johannes. Concerto no. 2 in B-flat, op. 83. Perf. Artur Rubinstein. Cond. Eugene Ormandy. Philadelphia Orch. RCA, 1992.

58. A published letter

Buttolph, Mrs. Laura E. Letter to Rev. and Mrs. C. C. Jones.

20 June 1857. In The Children of Pride: A True Story of Georgia and the Civil War. Ed. Robert Manson Myers. New Haven: Yale UP, 1972. 334.

59. A personal letter

Packer, Ann E. Letter to the author. 15 June 2004.

60. A lecture or address

Carlone, Dennis. "Architecture for the City of the Twenty-First Century." Symposium on the City. Urban Issues Group. Cambridge City Hall, Cambridge. 22 Oct. 2004.

61. An interview

Graaf, Vera. Personal interview. 19 Dec. 2004.

Rumsfeld, Donald. Interview. Frontline. PBS. WGBH, Boston. 13 Oct. 2004.

62. A map or other illustration

Women in the Armed Forces. Map. Women in the World: An International Atlas. By Joni Seager and Ann Olson. New York: Touchstone, 2004. 44-45.

36c ▪ MLA paper format

The document format recommended by the *MLA Handbook* is fairly simple, with just a few elements. See pages 161–62 for illustrations of the elements. See also pages 200–05 for guidelines on type fonts, headings, lists, illustrations, and other features that are not specified in MLA style.

Margins Use minimum one-inch margins on all sides of every page.

Spacing and indentions Double-space throughout. Indent paragraphs one-half inch or five spaces. (See the next page for indention of poetry and long prose quotations.)

Paging Begin numbering on the first page, and number consecutively through the end (including the list of works cited). Type Arabic numerals (1, 2, 3) in the upper right about one-half inch from the top. Place your last name before the page number in case the pages later become separated.

Identification and title In the upper left of the first page, give your name, your instructor's name, the course title, and the date—all double-spaced. Center the title. Do not type it in all-capitals or italics or place it between quotation marks.

Poetry and long prose quotations Treat a single line of poetry like any other quotation, running it into your text and enclosing it in quotation marks. You may run in two or three lines of poetry as well, separating the lines with a slash:

> An example of Robert Frost's incisiveness is in two lines from "Death of the Hired Man": "Home is the place where, when you have to go there / They have to take you in" (119-20).

Always separate quotations of more than three lines of poetry from your text. Use double spacing and a one-inch or ten-space indention. *Do not add quotation marks.*

> Emily Dickinson stripped ideas to their essence, as in this description of "A narrow Fellow in the Grass," a snake:
>
> I more than once at Noon
> Have passed, I thought, a Whip lash
> Unbraiding in the Sun
> When stopping to secure it
> It wrinkled, and was gone – (12-16)

MLA 36c

Also separate prose quotations of five or more typed lines. *Do not add quotation marks.*

> In his 1967 study of the lives of unemployed black men, Elliot Liebow observes that "unskilled" construction work requires more experience and skill than is generally assumed.
>
> A healthy, sturdy, active man of good intelligence requires from two to four weeks to break in on a construction job. . . . It frequently happens that his foreman or the craftsman he services is not willing to wait that long for him to get into condition or to learn at a glance the difference in size between a rough 2 x 8 and a finished 2 x 10. (62)

36d ▪ Sample pages in MLA style

First page

½″

Haley 1

All double-spaced

Vanessa Haley

Professor Moisan

English 101

24 March 2004

Annie Dillard's Healing Vision ← Center

½″ or 5 spaces

It is almost a commonplace these days that human arrogance is destroying the environment. Environmentalists, naturalists, and now the man or woman on the street seem to agree: the long-held belief that human beings are separate from nature, destined to rise above its laws and conquer it, has been ruinous.

1″ 1″

Unfortunately, much writing about nature goes to the opposite extreme, viewing nature as pure and harmonious, humanity as corrupt and dangerous. One nature writer who seems to recognize the naturalness of humanity is Annie Dillard. In her best-known work, the Pulitzer Prize-winning Pilgrim at Tinker Creek, she is a solitary person encountering the natural world, and some critics fault her for turning her back on society. But in those encounters with nature, Dillard probes a spiritual as well as a physical identity between human beings and nature that could help to heal the rift between them.

Dillard is not renowned for her sense of involvement with human society. Like Henry David Thoreau, with whom she is often compared, she retreats from rather than confronts human society. The critic Gary McIlroy points out that although Thoreau discusses society a great deal in Walden, he makes no attempt "to find a middle ground between it and his experiment in the woods" (113). Dillard has been similarly criticized. For instance, the writer Eudora Welty comments that

1″ or 10 spaces

> Annie Dillard is the only person in her book, substantially the only one in her world; I recall no outside human speech coming to break the long soliloquy of the author. Speaking of the universe very often, she is yet self-surrounded and, beyond that, book-surrounded. Her own book might have taken in more of human life without losing a bit of the wonder she was after. (37)

It is true, as Welty says, that in Pilgrim Dillard seems detached from human society. However, she actually was always

1″

MLA 36d

Second page

Haley 2

close to it at Tinker Creek. In a later book, Teaching a Stone to Talk, she says of the neighborhood, "This is, mind you, suburbia. It is a five-minute walk in three directions to rows of houses. . . . There's a 55 mph highway at one end of the pond, and a nesting pair of wood ducks at the other" (qtd. in Suh).

Rather than hiding from humanity, Dillard seems to be trying to understand it through nature. In Pilgrim she reports buying a goldfish, which she names Ellery Channing. She recalls once seeing through a microscope "red blood cells whip, one by one, through the capillaries" of yet another goldfish (124). Now watching Ellery Channing, she sees the blood in his body as a bond between fish and human being: "Those red blood cells are coursing in Ellery's tail now, too, in just that way, and through his mouth and eyes as well, and through mine" (125). Gary McIlroy observes that this blood, "a symbol of the sanctity of life, is a common bond between Dillard and the fish, between animal and human life in general, and between Dillard and other people" (115).

For Dillard, the terror and unpredictability of death unify all life. The most sinister image in Pilgrim--one that haunts Dillard--is that of the frog and the water bug. Dillard reports

Works cited

½″

Haley 6

Works Cited ← Center

All double-spaced

½″ or 5 spaces

1″

1″

Becker, John E. "Science and the Sacred: From Walden to Tinker Creek." Thought: A Review of Culture and Idea 62 (1987): 400-13.

Dillard, Annie. Pilgrim at Tinker Creek. New York: Harper, 1974.

Johnson, Sandra Humble. The Space Between: Literary Epiphany in the Work of Annie Dillard. Kent: Kent State UP, 1992.

McIlroy, Gary. "Pilgrim at Tinker Creek and the Social Legacy of Walden." South Atlantic Quarterly 85.2 (1986): 111-16.

Suh, Grace. "Ideas Are Tough, Irony Is Easy." Yale Herald Online 4 Oct. 2003. 22 Feb. 2004 <http://yaleherald.com/archive/xxii/10.4.03/ae/dillard.html>.

Welty, Eudora. Rev. of Pilgrim at Tinker Creek, by Annie Dillard. New York Times Book Review 24 Mar. 1974: 36-37. New York Times Ondisc. CD-ROM. UMI-ProQuest. June 1994.

37 APA Documentation and Format

The documentation style of the American Psychological Association is used in psychology and some other social sciences and is very similar to the styles in sociology, economics, and other disciplines. The following adapts the APA style from the *Publication Manual of the American Psychological Association*, 5th ed. (2001).

Note The APA provides occasional updates of its style and answers to frequently asked questions at *http://www.apastyle.org/faqs.html*.

37a ▪ APA parenthetical text citations

Citation formats

In APA style, parenthetical citations in the text refer to a list of sources at the end of the text. The basic parenthetical citation contains the author's last name and the date of publication. If you name the author in your text, then the citation includes just the date. Unless none is available, the APA also requires a page or other identifying number for a direct quotation and recommends an identifying number for a paraphrase.

APA PARENTHETICAL TEXT CITATIONS

Visit *www.ablongman.com/littlebrown* for added help and an exercise on APA documentation style.

1. Author not named in your text

One critic of Milgram's experiments insisted that the subjects "should have been fully informed of the possible effects on them" (Baumrind, 1968, p. 34).

2. Author named in your text

Baumrind (1968) insisted that the subjects in Milgram's study "should have been fully informed of the possible effects on them" (p. 34).

3. A work with two authors

Pepinsky and DeStefano (1997) demonstrated that a teacher's language often reveals hidden biases.

One study (Pepinsky & DeStefano, 1997) demonstrated hidden biases in a teacher's language.

4. A work with three to five authors

First reference:

Pepinsky, Dunn, Rentl, and Corson (1999) further demonstrated the biases evident in gestures.

Later references:

In the work of Pepinsky et al. (1999), the loaded gestures included head shakes and eye contact.

5. A work with six or more authors

One study (Rutter et al., 1996) attempted to explain these geographical differences in adolescent experience.

6. A work with a group author

An earlier prediction was even more somber (Lorenz Research, 2003).

Use this model for works that list an agency, institution, corporation, or other group as the author.

7. A work with no author or an anonymous work

One article ("Right to Die," 1996) noted that a death-row inmate may crave notoriety.

For a work that lists "anonymous" as the author, use this word in the citation: (Anonymous, 1999).

8. One of two or more works by the same author(s) published in the same year

At about age seven, most children begin to use appropriate gestures to reinforce their stories (Gardner, 1973a).

(See the reference for this source on p. 169.)

9. Two or more works by different authors

Two studies (Herskowitz, 1994; Marconi & Hamblen, 1999) found that periodic safety instruction can dramatically reduce employees' accidents.

10. An indirect source

Supporting data appear in a study by Wong (cited in Marconi & Hamblen, 1999).

11. An electronic source

Electronic sources can be cited like printed sources, usually with the author's last name and the publication date. When quoting or paraphrasing electronic sources that number paragraphs instead of pages, provide the paragraph number preceded by the symbol "¶" if you have it, or by "para."

Ferguson and Hawkins (1998) did not anticipate the "evident hostility" of participants (¶ 6).

If the source does not have numbering of any kind, provide just the author's name and the date.

Footnotes for supplementary content

When you need to explain something in your text—for instance, commenting on a source or providing data that don't fit into the relevant paragraph—you may place the supplementary information in a footnote. Follow the instructions for footnotes in the Chicago style (p. 179). Be careful not to overuse such notes: they can be more distracting than helpful.

37b ▪ APA reference list

In APA style, the in-text parenthetical citations refer to the list of sources at the end of the text. In this list, titled "References," you include full publication information on every source cited in your paper. The reference list falls at the end of the paper, numbered in sequence with the preceding pages. For an illustration of the following elements and their spacing, see page 178.

Arrangement Arrange sources alphabetically by the author's last name or, if there is no author, by the first main word of the title.

Spacing Double-space all entries.

Indention Begin each entry at the left margin, and indent the second and subsequent lines one-half inch or five to seven spaces.

Authors For works with up to six authors, list all authors with last name first, separating names and parts of names with commas. Use initials for first and middle names. Use an ampersand (&) before the last author's name. See model 3 on page 168 for treatment of seven or more authors.

Publication date Place the publication date in parentheses after the author's or authors' names, followed by a period. Generally, this date is the year only, though for some sources (such as magazine and newspaper articles) it includes month and sometimes day as well.

Title In the title of a book or article, capitalize only the first word of the title, the first word of the subtitle, and proper nouns;° all other words begin with small letters. In the title of a journal, capitalize all significant words. Italicize the titles of books and journals, along with any comma following. Do not italicize or use quotation marks around the titles of articles.

City of publication For print sources that are not periodicals (such as books or government publications), give the city of publication. With Baltimore, Boston, Chicago, Los Angeles, New York, Philadelphia, and San Francisco, follow the city name with a colon. With other cities, add a comma after the city name, give the two-letter postal abbreviation of the state, and then add a colon. (You may omit the state if the publisher is a university whose name includes the state name, such as "University of Arizona.")

Publisher's name For nonperiodical print sources, give the publisher's name after the place of publication and a colon. Use shortened names for many publishers (such as "Morrow" for William Morrow), and omit "Co.," "Inc.," and "Publishers." However, give full names for associations, corporations, and university presses, and do not omit "Books" or "Press" from a publisher's name.

Page numbers Use the abbreviation "p." or "pp." before page numbers in books and in newspapers, but *not* in

APA REFERENCES

other periodicals. For inclusive page numbers, include all figures: "667–668."

Punctuation Separate the parts of the reference (author, date, title, and publication information) with a period and one space. Do not use a final period in references to electronic sources, which conclude with an electronic address (see pp. 170–73).

Books

1. A book with one author

Rodriguez, R. (1982). *A hunger of memory: The education of Richard Rodriguez.* Boston: Godine.

2. A book with two to six authors

Nesselroade, J. R., & Baltes, P. B. (1999). *Longitudinal research in the study of behavioral development.* New York: Academic Press.

3. A book with seven or more authors

Wimple, P. B., Van Eijk, M., Potts, C. A., Hayes, J., Obergau, W. R., Zimmer, S., et al. (2001). *Case studies in moral decision making among adolescents.* San Francisco: Jossey-Bass.

4. A book with an editor

Dohrenwend, B. S., & Dohrenwend, B. P. (Eds.). (1999). *Stressful life events: Their nature and effects.* New York: Wiley.

5. A book with a translator

Trajan, P. D. (1927). *Psychology of animals.* (H. Simone, Trans.). Washington, DC: Halperin.

6. A book with a group author

Lorenz Research, Inc. (2005). *Research in social studies teaching*. Baltimore: Arrow Books.

7. A book with no author or an anonymous book

Merriam-Webster's collegiate dictionary (11th ed.). (2003). Springfield, MA: Merriam-Webster.

For a work whose author is actually given as "Anonymous," use this word in place of the author's name and alphabetize it as if it were a name.

8. Two or more works by the same author(s) published in the same year

Gardner, H. (1973a). *The arts and human development.* New York: Wiley.

Gardner, H. (1973b). *The quest for mind: Piaget, Lévi-Strauss, and the structuralist movement.* New York: Knopf.

9. A later edition

Bolinger, D. L. (1975). *Aspects of language* (2nd ed.). New York: Harcourt Brace Jovanovich.

10. A work in more than one volume

Reference to a single volume:

Lincoln, A. (1953). *The collected works of Abraham Lincoln* (R. P. Basler, Ed.). (Vol. 5). New Brunswick, NJ: Rutgers University Press.

Reference to all volumes:

Lincoln, A. (1953). *The collected works of Abraham Lincoln* (R. P. Basler, Ed.). (Vols. 1-8). New Brunswick, NJ: Rutgers University Press.

11. An article or chapter in an edited book

Paykel, E. S. (1999). Life stress and psychiatric disorder: Applications of the clinical approach. In B. S. Dohrenwend & B. P. Dohrenwend (Eds.), *Stressful life events: Their nature and effects* (pp. 239-264). New York: Wiley.

Periodicals: Journals, magazines, newspapers

12. An article in a journal with continuous pagination throughout the annual volume

Emery, R. E. (2003). Marital turmoil: Interpersonal conflict and the children of divorce. *Psychological Bulletin, 92,* 310-330.

13. An article in a journal that pages issues separately

Dacey, J. (1998). Management participation in corporate buy-outs. *Management Perspectives,* 7(4), 20-31.

14. An abstract of a journal article

Emery, R. E. (2003). Marital turmoil: Interpersonal conflict and the children of discord and divorce. *Psychological Bulletin, 92,* 310-330. Abstract obtained from *Psychological Abstracts,* 1992, *69,* Item 1320.

15. An article in a magazine

Talbot, M. (2004, August 9). The bad mother. *The New Yorker,* 40-46.

16. An article in a newspaper

Kolata, G. (2004, January 7). Kill all the bacteria! *The New York Times,* pp. B1, B3.

17. An unsigned article

The right to die. (1996, October 11). *Time, 121,* 101.

18. A review

Dinnage, R. (1987, November 29). Against the master and his men [Review of the book *A mind of her own: The life of Karen Horney*]. *The New York Times Book Review,* 10-11.

Electronic sources

In general, the APA's electronic-source references begin as those for print references do: author(s), date, title. Then you add information on when and how you retrieved the source. For example, an online source might conclude Retrieved August 6, 2004, from http://sites.unc.edu/~daniel/social_constructionism (in APA style, no period follows a URL at the end of the reference). When you need to divide a URL from one line to the next, APA style calls for breaking *only* after a slash or before a period. Do not hyphenate a URL.

Try to locate all the information required by a model, referring to pages 110–11 for help. However, if you search for and still cannot find some information, then give what you can find. If a source has no publication date, use "n.d." (for *no date*) in place of a publication date (see model 28, p. 172).

19. A journal article published online and in print

Palfrey, A. (1996). Choice of mates in identical twins [Electronic version]. *Modern Psychology, 4*(1), 26-40.

20. An article in an online journal

Wissink, J. A. (2000). Techniques of smoking cessation among teens and adults. *Adolescent Medicine, 2.* Retrieved August 16, 2004, from http://www.easu.edu/AdolescentMedicine/2-Wissink.html

21. A journal article from an electronic database

Wilkins, J. M. (1999). The myths of the only child. *Psychology Update 11*(1), 16-23. Retrieved December 20, 2004, from ProQuest Direct database.

22. An abstract from an electronic database

Wilkins, J. M. (1999). The myths of the only child. *Psychology Update, 11*(1), 16-23. Abstract retrieved December 20, 2004, from ProQuest Direct database.

23. An article in an online newspaper

Pear, R. (2004, January 23). Gains reported for children of welfare to work families. *The New York Times on the Web.* Retrieved January 19, 2005, from http://www.nytimes.com/2004/01/23/national/23/WELF.html

24. An entire web site (text citation)

The APA's Web site provides answers to frequently asked questions about style (http://www.apa.org).

Cite an entire Web site (rather than a specific page or document) by giving the electronic address in your text.

25. An independent document on the Web

Anderson, D. (2004, May 1). *Social constructionism and MOOs.* Retrieved August 6, 2004, from http://sites.unc.edu/~daniel/social_constructionism

26. A document from the Web site of a university or government agency

McConnell, L. M., Koenig, B. A., Greely, H. T., & Raffin, T. A. (2004, August 17). *Genetic testing and Alzheimer's disease: Has the time come?* Retrieved September 1, 2004, from Stanford University, Project in Genomics, Ethics, and Society Web site: http://scbe.stanford.edu/pges

27. An online government report

U.S. Department of Commerce. National Telecommunications and Information Administration. (2001, October). *Falling through the net: Toward digital inclusion.* Retrieved April 12, 2004, from http://www.ntia.doc.gov/ntiahome/digitaldivide

28. A multipage online document

Elston, C. (n.d.). *Multiple intelligences.* Retrieved June 6, 2004, from http://education.com/teachspace/intelligences

29. A part of an online document

Elston, C. (n.d.). Logical/math intelligence. In *Multiple intelligences.* Retrieved June 6, 2004, from http://education.com/teachspace/intelligences/logical.jsp

30. A retrievable online posting

Tourville, M. (2005, January 6). European currency reform. Message posted to International Finance electronic mailing list, archived at http://www.liasu.edu/finance-dl/46732

Include postings to discussion lists and newsgroups in your list of references *only* if they are retrievable by others. The source above is archived (as the reference makes plain) and thus retrievable at the address given.

31. Electronic mail or a nonretrievable online posting (text citation)

At least one member of the research team has expressed reservations about the design of the study (L. Kogod, personal communication, February 6, 2005).

Personal electronic mail and other online postings that are not retrievable by others should be cited only in your text, as in the example above.

32. Computer software

Project scheduler 9000 [Computer software]. (2005). Orlando, FL: Scitor.

Other sources

33. A report

Gerald, K. (1998). *Micro-moral problems in obstetric care* (Report No. NP-71). St. Louis: Catholic Hospital Association.

Jolson, M. K. (1991). *Music education for preschoolers* (Report No. TC-622). New York: Teachers College, Columbia University. (ERIC Document Reproduction Service No. ED 264 488)

34. A government publication

Hawaii. Department of Education. (2005). *Kauai district schools, profile 2003-04.* Honolulu, HI: Author.

Stiller, A. (1996). *Historic preservation and tax incentives.* Washington, DC: U.S. Department of the Interior.

U.S. House. Committee on Ways and Means. (2003). *Medicare payment for outpatient physical and occupational therapy services.* 108th Cong., 1st Sess. Washington, DC: U.S. Government Printing Office.

35. A doctoral dissertation

A dissertation abstracted in *DAI* and obtained from UMI:

Steciw, S. K. (1986). Alterations to the Pessac project of Le Corbusier. *Dissertation Abstracts International, 46,* 565C. (UMI No. 6216202)

A dissertation abstracted in *DAI* and obtained from the university:

Chang, J. K. (2000). Therapeutic intervention in treatment of injuries to the hand and wrist (Doctoral dissertation, University of Michigan, 2000). *Dissertation Abstracts International, 50,* 162.

An unpublished dissertation:

Delaune, M. L. (2001). *Child care in single-mother and single-father families: Differences in time, activity, and stress.* Unpublished doctoral dissertation, University of California, Davis.

36. A published interview

Brisick, W. C. (1998, July 1). [Interview with Ishmael Reed]. *Publishers Weekly,* 41-42.

For an interview you conduct yourself, use an in-text parenthetical citation, as shown in model 31 (previous page) for a nonretrievable online posting.

37. A motion picture

Spielberg, S. (Director). (1993). *Schindler's list* [Motion picture]. United States: Viacom.

American Psychological Association (Producer). (2001). *Ethnocultural psychotherapy* [Motion picture]. (Available from the American Psychological Association, 750 First Street, NE, Washington, DC 20002-4242, or online from http://www.apa.org/videos/4310240.html)

38. A musical recording

Springsteen, B. (2002). Empty sky. *The rising* [CD]. New York: Columbia.

39. A television series or episode

Taylor, C., Cleveland, R., & Andries, L. (Producers). (2004). *Six feet under* [Television series]. New York: HBO.

Cleveland, R. (writer), & Engler, M. (Director). (2004). Dillon Michael Cooper [Television series episode]. In C. Taylor,

R. Cleveland, & L. Andries (Producers), *Six feet under.* New York: HBO.

37c ▪ APA paper format

The APA *Publication Manual* distinguishes between documents intended for publication (which will be set in type) and those submitted by students (which are the final copy). The guidelines below apply to most undergraduate papers. Check with your instructor for any modifications to this format.

For illustrations of the following elements, see pages 177–78. And see pages 200–05 for guidelines on type fonts, lists, tables and figures, and other elements of document design.

Margins Use one-inch margins on the top, bottom, and right side. Add another half-inch on the left to accommodate a binder.

Spacing and indentions Double-space your text and references. (See the next page for spacing of displayed quotations.) Indent paragraphs and displayed quotations one-half inch or five to seven spaces.

Paging Begin numbering on the title page, and number consecutively through the end (including the reference list). Type Arabic numerals (1, 2, 3) in the upper right, about one-half inch from the top.

Place a shortened version of your title five spaces to the left of the page number.

Title page Include the full title, your name, the course title, the instructor's name, and the date. Type the title on the top half of the page, followed by the identifying information, all centered horizontally and double-spaced. Include a shortened form of the title along with the page number at the top of this and all other pages.

Abstract Summarize (in a maximum of 120 words) your subject, research method, findings, and conclusions. Put the abstract on a page by itself.

Body Begin with a restatement of the paper's title and then an introduction (not labeled). The introduction concisely presents the problem you researched, your research method, the relevant background (such as related studies), and the purpose of your research.

The "Method" section provides a detailed discussion of how you conducted your research, including a description of the research subjects, any materials or tools you used (such as questionnaires), and the procedure you followed.

The "Results" section summarizes the data you collected, explains how you analyzed them, and presents them in detail, often in tables, graphs, or charts.

The "Discussion" section interprets the data and presents your conclusions. (When the discussion is brief, you may combine it with the previous section under the heading "Results and Discussion.")

Headings Label the "Method," "Results," and "Discussion" sections with centered first-level headings, and use second- and third-level headings as needed. Double-space all headings.

First-Level Heading

Second-Level Heading

Third-level heading. Run this heading into the text paragraph.

Long quotations Run into your text all quotations of forty words or less, and enclose them in quotation marks. For quotations of more than forty words, set them off from your text by indenting all lines one-half inch or five to seven spaces, double-spacing above and below. For student papers, the APA allows single-spacing of displayed quotations:

Echoing the opinions of other Europeans at the time, Freud (1961) had a poor view of Americans:

> The Americans are really too bad. . . . Competition is much more pungent with them, not succeeding means civil death to every one, and they have no private resources apart from their profession, no hobby, games, love or other interests of a cultured person. And success means money. (p. 86)

Do not use quotation marks around a quotation displayed in this way.

37d ▪ Sample pages in APA style

Title page

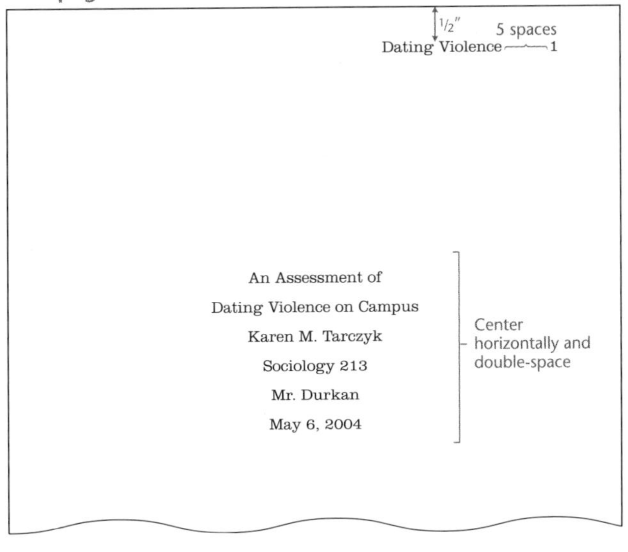

Abstract

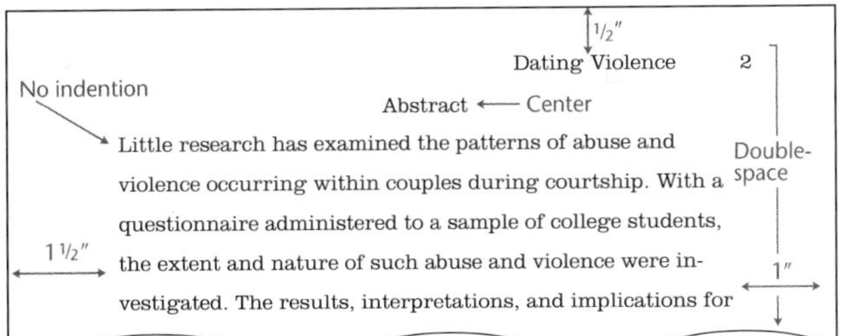

First page of body

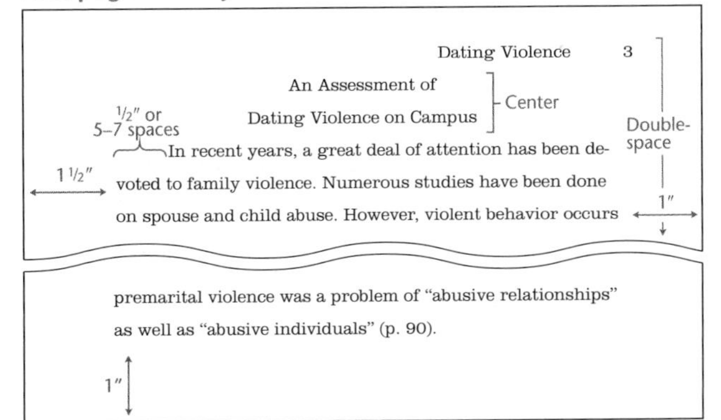

Later page of body

Dating Violence 4

All the studies indicate a problem that is being neglected. My objective was to gather data on the extent and nature of premarital violence and to discuss possible interpretations.

Method

Sample

I conducted a survey of 200 students (134 females, 66 males) at a large state university in the northeastern United States. The sample consisted of students enrolled in an introductory sociology course.

References

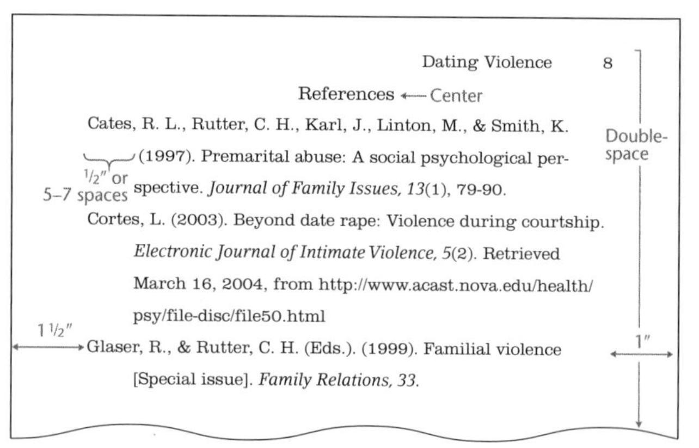

Dating Violence 8

References

Cates, R. L., Rutter, C. H., Karl, J., Linton, M., & Smith, K. (1997). Premarital abuse: A social psychological perspective. *Journal of Family Issues, 13*(1), 79-90.

Cortes, L. (2003). Beyond date rape: Violence during courtship. *Electronic Journal of Intimate Violence, 5*(2). Retrieved March 16, 2004, from http://www.acast.nova.edu/health/psy/file-disc/file50.html

Glaser, R., & Rutter, C. H. (Eds.). (1999). Familial violence [Special issue]. *Family Relations, 33.*

38 Chicago Documentation and Format

Writers in history, art history, philosophy, and other humanities use the note style of documentation from *The Chicago Manual of Style*, 15th ed. (2003), or the student reference adapted from its previous edition, *A Manual for Writers of Term Papers, Theses, and Dissertations,* by Kate L. Turabian, 6th ed., revised by John Grossman and Alice Bennett (1996). *The Chicago Manual* has a Web site that

Visit *www.ablongman.com/littlebrown* for added help with Chicago documentation style.

answers frequently asked questions about its style: *http://www.press.uchicago.edu/Misc/Chicago/cmosfaq/cmosfaq.html.*

This chapter explains the Chicago note style (below) and Chicago paper format (p. 189).

38a ▪ Chicago notes and list of works cited

In the Chicago note style, a raised numeral in the text refers the reader to source information in endnotes or footnotes. In these notes, the first citation of each source contains all the information readers need to find the source. Thus your instructor may consider a list of works cited optional because it provides much the same information. Ask your instructor whether you should use footnotes or endnotes and whether you should include a list of works cited.

The examples below illustrate the essentials of a note and a works-cited entry. (See p. 191 for additional illustrations.)

Note

6. Martin Gilbert, *Atlas of British History* (New York: Dorset, 1988), 96.

Works-cited entry

Gilbert, Martin. *Atlas of British History*. New York: Dorset, 1988.

Notes and works-cited entries share certain features:

- Single-space each note or entry, and double-space between them.
- Italicize or underline the titles of books and periodicals (ask your instructor for his or her preference).
- Enclose in quotation marks the titles of parts of books or articles in periodicals.
- Do not abbreviate publishers' names, but omit "Inc.," "Co.," and similar abbreviations.
- Do not use "p." or "pp." before page numbers.

Notes and works-cited entries also differ in important ways.

Note	Works-cited entry
Start with a number that corresponds to the note number in the text.	Do not begin with a number.
Indent the first line five spaces.	Indent the second and subsequent lines five spaces.
Give the author's name in normal order.	Begin with the author's last name.

Use commas between elements such as author's name and title.	Use periods between elements, followed by one space.
Enclose publication information in parentheses, with no preceding punctuation.	Precede the publication information with a period, and don't use parentheses.
Include the specific page number(s) you borrowed from, omitting "p." or "pp."	Omit page numbers except for parts of books or articles in periodicals.

Many word-processing programs automatically position footnotes at the bottoms of appropriate pages. Some automatically number notes and even renumber them if you add or delete one or more.

38b ▪ Chicago models

In the following models for common sources, notes and works-cited entries appear together for easy reference. In your papers, be sure to use the numbered note form for notes and the unnumbered works-cited form for works-cited entries.

Books

1. A book with one, two, or three authors

1. Carol Gilligan, *In a Different Voice: Psychological Theory and Women's Development* (Cambridge: Harvard University Press, 1982), 27.

Gilligan, Carol. *In a Different Voice: Psychological Theory and Women's Development.* Cambridge: Harvard University Press, 1982.

1. Dennis L. Wilcox, Phillip H. Ault, and Warren K. Agee, *Public Relations: Strategies and Tactics,* 4th ed. (New York: HarperCollins, 1995), 182.

Wilcox, Dennis L., Phillip H. Ault, and Warren K. Agee. *Public Relations: Strategies and Tactics.* 4th ed. New York: HarperCollins, 1995.

2. A book with more than three authors

2. Geraldo Lopez and others, *China and the West* (Boston: Little, Brown, 2000), 461.

Lopez, Geraldo, Judith P. Salt, Anne Ming, and Henry Reisen. *China and the West.* Boston: Little, Brown, 2000.

CHICAGO NOTE AND WORKS-CITED MODELS

3. A book with an editor

3. Hendrick Ruitenbeek, ed., *Freud as We Knew Him* (Detroit: Wayne State University Press, 1973), 64.

Ruitenbeek, Hendrick, ed. *Freud as We Knew Him.* Detroit: Wayne State University Press, 1973.

4. A book with an author and an editor

4. Lewis Mumford, *The City in History,* ed. Donald L. Miller (New York: Pantheon, 1986), 216-17.

Mumford, Lewis. *The City in History.* Edited by Donald L. Miller. New York: Pantheon, 1986.

5. A translation

5. Dante Alighieri, *The Inferno,* trans. John Ciardi (New York: New American Library, 1971), 51.

Alighieri, Dante. *The Inferno.* Translated by John Ciardi. New York: New American Library, 1971.

6. An anonymous work

6. *The Dorling Kindersley World Reference Atlas* (London: Dorling Kindersley, 2005), 150-51.

The Dorling Kindersley World Reference Atlas. London: Dorling Kindersley, 2005.

7. A later edition

7. Dwight L. Bolinger, *Aspects of Language,* 2nd ed. (New York: Harcourt Brace Jovanovich, 1975), 20.

Bolinger, Dwight L. *Aspects of Language.* 2nd ed. New York: Harcourt Brace Jovanovich, 1975.

8. A work in more than one volume

Citation of one volume without a title:

8. Abraham Lincoln, *The Collected Works of Abraham Lincoln,* ed. Roy P. Basler (New Brunswick: Rutgers University Press, 1953), 5:426-28.

Lincoln, Abraham. *The Collected Works of Abraham Lincoln.* Edited by Roy P. Basler. Vol. 5. New Brunswick: Rutgers University Press, 1953.

Citation of one volume with a title:

8. Linda B. Welkin, *The Age of Balanchine,* vol. 3 of *The History of Ballet* (New York: Columbia University Press, 1969), 56.

Welkin, Linda B. *The Age of Balanchine.* Vol. 3 of *The History of Ballet.* New York: Columbia University Press, 1969.

9. A selection from an anthology

9. Rosetta Brooks, "Streetwise," in *The New Urban Landscape,* ed. Richard Martin (New York: Rizzoli, 2001), 38-39.

Brooks, Rosetta. "Streetwise." In *The New Urban Landscape,* edited by Richard Martin, 37-60. New York: Rizzoli, 2001.

10. A work in a series

10. Ingmar Bergman, *The Seventh Seal,* Modern Film Scripts, no. 12 (New York: Simon and Schuster, 1968), 27.

Bergman, Ingmar. *The Seventh Seal.* Modern Film Scripts, no. 12. New York: Simon and Schuster, 1968.

11. An article in a reference work

Chicago does not require publication information for well-known reference works like the one listed here. The abbreviation "s.v." stands for the Latin *sub verbo,* "under the word."

11. *Merriam-Webster's Collegiate Dictionary,* 11th ed., s.v. "reckon."

Merriam-Webster's Collegiate Dictionary, 11th ed., s.v. "reckon."

Periodicals: Journals, magazines, newspapers

12. An article in a journal

12. Janet Lever, "Sex Differences in the Games Children Play," *Social Problems* 23 (1996): 482.

Lever, Janet. "Sex Differences in the Games Children Play." *Social Problems* 23 (1996): 478-87.

12. June Dacey, "Management Participation in Corporate Buy-Outs," *Management Perspectives* 7, no. 4 (1998): 22.

Dacey, June. "Management Participation in Corporate Buy-Outs." *Management Perspectives* 7, no. 4 (1998): 20-31.

13. An article in a magazine

13. Margaret Talbot, "The Bad Mother," *New Yorker,* August 9, 2004, 32.

Talbot, Margaret. "The Bad Mother." *New Yorker,* August 9, 2004, 31-37.

14. An article in a newspaper

14. Gina Kolata, "Kill All the Bacteria!" *New York Times,* January 7, 2005, national edition, B1.

Kolata, Gina. "Kill All the Bacteria!" *New York Times,* January 7, 2005, national edition, B1, B6.

15. A review

15. John Gregory Dunne, "The Secret of Danny Santiago," review of *Famous All over Town,* by Danny Santiago, *New York Review of Books,* August 16, 1994, 25.

Dunne, John Gregory. "The Secret of Danny Santiago." Review of *Famous All over Town,* by Danny Santiago. *New York Review of Books,* August 16, 1994, 17-27.

Electronic sources

The Chicago Manual's models for documenting electronic sources derive mainly from those for print sources, with the addition of an electronic address (URL) or other indication of the medium along with any other information that may help readers locate the source. Chicago requires the date of your access to an online source only if the source could change significantly (for instance, a report on medical research). However, your instructor may require access dates for a broader range of online sources, so they are included in the following models (in parentheses at the end).

Note Chicago style allows many ways to break URLs between the end of one line and the beginning of the next: after slashes, before most punctuation marks (periods, commas, question marks, and so on), and before or after equal signs and ampersands (&). *Do not* break after a hyphen or add any hyphens.

16. A work on CD-ROM or DVD-ROM

16. *The American Heritage Dictionary of the English Language,* 4th ed., CD-ROM (Boston: Houghton Mifflin, 2000).

The American Heritage Dictionary of the English Language. 4th ed. CD-ROM. Boston: Houghton Mifflin, 2000.

17. A work from an online database

17. Irina Netchaeva, "E-Government and E-Democracy." *International Journal for Communication Studies* 64 (2002): 470-71, http://www.epnet.com (accessed December 20, 2004).

Netchaeva, Irina. "E-Government and E-Democracy." *International Journal for Communication Studies* 64 (2002): 467-78. http://www.epnet.com (accessed December 20, 2004).

For news and journal databases, including those to which your library subscribes, you may omit the name of the database. Give its main URL (as in the examples) unless the work has a usable URL of its own. (See pp. 155–56 for more on database URLs.)

18. An online book

18. Jane Austen, *Emma,* ed. Ronald Blythe (1816; Harmondsworth, UK: Penguin, 1972; Oxford Text Archive, 2000), chap. 1, http://ota.ox.ac.uk/public/english/Austen/emma.1519 (accessed December 15, 2004).

Austen, Jane. *Emma.* Edited by Ronald Blythe. 1816. Harmondsworth, UK: Penguin, 1972. Oxford Text Archive, 2000. http:// ota.ox.ac.uk/public/english/Austen/emma.1519 (accessed December 15, 2004).

19. An article in an online journal

19. Andrew Palfrey, "Choice of Mates in Identical Twins," *Modern Psychology* 4, no. 1 (1996): 28, http://www.liasu/edu/modpsy/palfrey4(1).htm (accessed February 25, 2005).

Palfrey, Andrew. "Choice of Mates in Identical Twins." *Modern Psychology* 4, no. 1 (1996): 26-40. http://www.liasu/edu/modpsy/palfrey4(1).htm (accessed February 25, 2005).

20. An article in an online magazine

20. Ricki Lewis, "The Return of Thalidomide," *Scientist,* January 22, 2001, http://www.the-scientist.com/yr2001/jan/lewis_pl_010122.html (accessed January 24, 2005).

Lewis, Ricki. "The Return of Thalidomide. *Scientist,* January 22, 2001. http://www.the-scientist.com/yr2001/jan/lewis_pl_010122.html (accessed January 24, 2005).

21. An article in an online newspaper

21. Lucia Still, "On the Battlefields of Business, Millions of Casualties," *New York Times on the Web,* March 3, 2002, http://www.nytimes.com/specials/downsize/03down1.html (accessed August 17, 2004).

Still, Lucia. "On the Battlefields of Business, Millions of Casualties." *New York Times on the Web,* March 3, 2002. http://www.nytimes.com/specials/downsize/03down1.html (accessed August 17, 2004).

22. An article in an online reference work

22. *Encyclopaedia Britannica Online,* s.v. "Wu-ti," http://www.eb.com:80 (accessed December 23, 2004).

Encyclopaedia Britannica Online. S.v. "Wu-ti." http://www.eb.com:80 (accessed December 23, 2004).

23. An online audio or visual source

A work of art:

23. Jackson Pollock, *Shimmering Substance,* oil on canvas, 1946, Museum of Modern Art, New York, WebMuseum, http://www.ibiblio.org/wm/paint/auth/Pollock/pollock.shimmering.jpg (accessed March 12, 2004).

Pollock, Jackson. *Shimmering Substance.* Oil on canvas, 1946. Museum of Modern Art, New York. WebMuseum. http://www.ibiblio.org/wm/paint/auth/Pollock/pollock.shimmering.jpg (accessed March 12, 2004).

A sound recording:

23. Ronald W. Reagan, State of the Union address, January 26, 1982, Vincent Voice Library, Digital and Multimedia Center, University of Michigan, http://www.lib.msu.edu/vincent/presidents/reagan.html (accessed May 6, 2004).

Reagan, Ronald W. State of the Union address. January 26, 1982. Vincent Voice Library. Digital and Multimedia Center, University of Michigan. http://www.lib.msu.edu/vincent/presidents/reagan.html (accessed May 6, 2004).

A film or film clip:

23. Leslie J. Stewart, *96 Ranch Rodeo and Barbecue* (1951), 16mm, from Library of Congress, *Buckaroos in Paradise: Ranching Culture in Northern Nevada, 1945-1982,* MPEG, http://memory/loc.gov/cgi-bin/query/ammem/ncr:@field(DocID+@lit(nv034)) (accessed January 7, 2005).

Stewart, Leslie J. 96 *Ranch Rodeo and Barbecue.* 1951, 16 mm. From Library of Congress, *Buckaroos in Paradise: Ranching Culture in Northern Nevada, 1945-1982.* MPEG, http://memory/loc.gov/cgi-bin/query/ammem/ncr:@field(DocID+@lit(nv034)) (accessed January 7, 2005).

24. A posting to an online discussion group

24. Michael Tourville, "European Currency Reform," e-mail to International Finance Discussion List, January 6, 2004, http://www.weg.isu/finance-dl/archive/46732 (accessed January 12, 2004).

Tourville, Michael. "European Currency Reform." E-mail to International Finance Discussion List. January 6, 2004. http://www.weg.isu/finance-dl/archive/46732 (accessed January 12, 2004).

25. Electronic mail

25. Michele Millon, "Re: Grief Therapy," e-mail message to author, May 4, 2004.

Millon, Michele. "Re: Grief Therapy." E-mail message to author. May 4, 2004.

Other sources

26. A government publication

26. House Committee on Ways and Means, *Medicare Payment for Outpatient Physical and Occupational Therapy Services,* 108th Cong., 1st sess., 2003, H. Doc. 409, 12-13.

U.S. Congress. House. Committee on Ways and Means. *Medicare Payment for Outpatient Physical and Occupational Therapy Services.* 108th Cong., 1st sess., 2003. H. Doc. 409.

26. Hawaii Department of Education, *Kauai District Schools, Profile 2003-04* (Honolulu, 2005), 27.

Hawaii. Department of Education. *Kauai District Schools, Profile 2003-04.* Honolulu, 2005.

27. A published letter

27. Mrs. Laura E. Buttolph to Rev. and Mrs. C. C. Jones, June 20, 1857, in *The Children of Pride: A True Story of Georgia and the Civil War,* ed. Robert Manson Myers (New Haven, CT: Yale University Press, 1972), 334.

Buttolph, Laura E. Mrs. Laura E. Buttolph to Rev. and Mrs. C. C. Jones, June 20, 1857. In *The Children of Pride: A True Story of Georgia and the Civil War,* edited by Robert Manson Myers. New Haven, CT: Yale University Press, 1972.

28. A published or broadcast interview

28. Donald Rumsfeld, interview by William Lindon, *Frontline,* PBS, October 13, 2004.

Rumsfeld, Donald. Interview by William Lindon. *Frontline.* PBS, October 13, 2004.

29. A personal letter or interview

29. Ann E. Packer, letter to author, June 15, 2004.

Packer, Ann E. Letter to author. June 15, 2004.

29. Vera Graaf, interview by author, December 19, 2004.

Graaf, Vera. Interview by author. December 19, 2004.

30. A work of art

30. John Singer Sargent, *In Switzerland,* watercolor, 1908, Metropolitan Museum of Art, New York.

Sargent, John Singer. *In Switzerland.* Watercolor, 1908. Metropolitan Museum of Art, New York.

31. A film or video recording

31. George Balanchine, *Serenade,* VHS, San Francisco Ballet (New York: PBS Video, 1985).

Balanchine, George. *Serenade.* VHS. San Francisco Ballet. New York: PBS Video, 1985.

32. A sound recording

32. Johannes Brahms, Piano Concerto no. 2 in B-flat, Artur Rubinstein, Philadelphia Orchestra, Eugene Ormandy, compact disc, RCA BRC4-6731.

Brahms, Johannes. Piano Concerto no. 2 in B-flat. Artur Rubinstein. Philadelphia Orchestra. Eugene Ormandy. Compact disc. RCA BRC4-6731.

Shortened notes

To streamline documentation, Chicago style recommends shortened notes for sources that are fully cited elsewhere, either in a complete list of works cited or in previous notes. Ask your instructor whether your paper should include a list of works cited and, if so, whether you may use shortened notes for first references to sources as well as for subsequent references.

A shortened note contains the author's last name, the work's title (minus any initial *A*, *An*, or *The*), and the page number. Reduce long titles to four or fewer key words.

Complete note

8. Janet Lever, "Sex Differences in the Games Children Play," *Social Problems* 23 (1996): 482.

Complete works-cited entry

Lever, Janet. "Sex Differences in the Games Children Play." *Social Problems* 23 (1996): 478-87.

Shortened note

12. Lever, "Sex Differences," 483.

You may use the Latin abbreviation "ibid." (meaning "in the same place") to refer to the same source cited in the preceding note. Give a page number if it differs from that in the preceding note.

12. Lever, "Sex Differences," 483.

13. Gilligan, *In a Different Voice,* 92.

14. Ibid., 93.

15. Lever, "Sex Differences," 483.

Chicago style allows for in-text parenthetical citations when you cite one or more works repeatedly. In the following example, the raised number 2 refers to the source information in a note; the number in parentheses is a page number in the same source.

> British rule, observes Stuart Cary Welch, "seemed as permanent as Mount Everest."[2] Most Indians submitted, willingly or not, to British influence in every facet of life (42).

38c ▪ Chicago paper format

The following guidelines come mainly from Turabian's *Manual for Writers,* which offers more specific advice than *The Chicago Manual* on the format of students' papers. See the next two pages for illustrations of the following elements. And see pages 200–05 for advice on type fonts, lists, illustrations, and other elements of document design.

Margins and spacing Use minimum one-inch margins on all pages of the body. (The first page of endnotes or works cited begins two inches from the top; see p. 191.) Double-space your own text and between notes and works-cited entries; single-space displayed quotations (see below) and each note and works-cited entry.

Paging Number pages consecutively from the first text page through the end (endnotes or works cited). Use Arabic numerals (1, 2, 3) in the upper right corner.

Title page On an unnumbered title page provide the title of the paper, your name, the course title, your instructor's name, and the date. Use all-capital letters, and center everything horizontally. Double-space between adjacent lines, and add extra space between elements as shown on the next page.

Poetry and long prose quotations Display certain quotations separately from your text: three or more lines of poetry and two or more sentences of prose. Indent a displayed quotation four spaces from the left, single-space the quotation, and double-space above and below it. *Do not add quotation marks.*

> Gandhi articulated the principles of his movement in 1922:
>
> > I discovered that pursuit of truth did not permit violence being inflicted on one's opponent, but that he must be weaned from error by patience and sympathy. For what appears to be truth to one may appear to be error to the other. And patience means self-suffering.[9]

38d ▪ Sample pages in Chicago style

Title page

INDIAN NATIONALISM IN INDIAN ART

← Double-space

AFTER WORLD WAR I

REYNA P. DIXON

ART HISTORY 236

MS. PARIKH — Double-space

DECEMBER 16, 2004

First page of paper with footnotes

1″ 1

1″ World War I was a transformative event for every partici- 1″
pant, not least for faraway India. Though at first unified around

In 1901, Madras, Bengal, and Punjab were a few of the huge Indian provinces governed by the British viceroy.[1] British rule, observes Stuart Cary Welch, "seemed as permanent as Mount 1″ Everest."[2] Most Indians submitted, willingly or not, to British 1″

———— ← Line

5 spaces

1. Martin Gilbert, *Atlas of British History* (New York: Dorset, 1968), 96. — Single-space

← Double-space

2. Stuart Cary Welch, *India: Art and Culture* (New York: Metropolitan Museum of Art, 1995), 42. — Single-space

1″

Endnotes

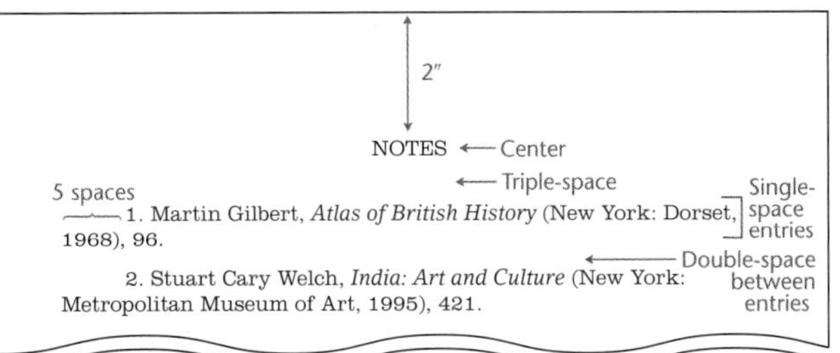

Bibliography

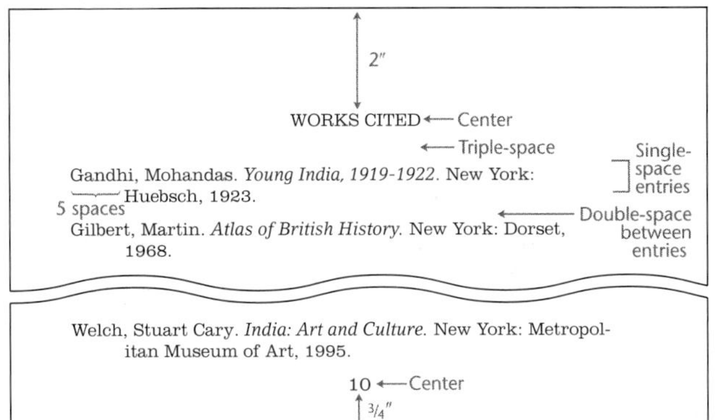

39 CSE Documentation

Writers in the life sciences, physical sciences, and mathematics rely for documentation style on *Scientific Style and Format: The CBE Style Manual for Authors, Editors, and Publishers* (6th ed., 1994). Its sponsoring organization, the Council of Science Editors, was until 2000 called the Council of Biology Editors, so you will see the style abbreviated both CSE (as here) and CBE.

Visit ***www.ablongman.com/littlebrown*** for added help with CSE documentation style.

Scientific Style and Format details both styles of scientific documentation: one using author and date and one using numbers. Both types of text citation refer to a list of references at the end of the paper (opposite). Ask your instructor which style you should use.

39a ▪ CSE name-year text citations

In the CSE name-year style, parenthetical text citations provide the last name of the author being cited and the source's year of publication. At the end of the paper, a list of references, arranged alphabetically by authors' last names, provides complete information on each source. (See opposite.)

The CSE name-year style closely resembles the APA name-year style detailed on pages 163–65. You can follow the APA examples for in-text citations, making several notable changes for CSE:

- Do not use a comma to separate the author's name and the date: (Baumrind 1968, p. 34).
- Separate two authors' names with "and" (not "&"): (Pepinsky and DeStefano 1997).
- Use "and others" (not "et al.") for three or more authors: (Rutter and others 1996).
- List unnamed or anonymous authors as "Anonymous," both in the citation and in the list of references: (Anonymous 1976).

39b ▪ CSE numbered text citations

In the CSE number style, raised numbers in the text refer to a numbered list of references at the end of the paper.

> Two standard references[1,2] use this term.
>
> These forms of immunity have been extensively researched.[3]

Assignment of numbers The number for each source is based on the order in which you cite the source in the text: the first cited source is 1, the second is 2, and so on.

Reuse of numbers When you cite a source you have already cited and numbered, use the original number again. This reuse is the key difference between the CSE numbered citations and numbered references to footnotes or endnotes.

Citation of two or more sources When you cite two or more sources at once, arrange their numbers in sequence and separate them with a comma and no space, as in the first of the preceding examples.

39c ▪ CSE reference list

For both the name-year and the number styles of in-text citation, provide a list, titled "References," of all sources you have cited. Format the page as shown for APA references on page 178 (but you may omit the shortened title before the page number).

Spacing Single-space each entry, and double-space between entries.

Arrangement In the name-year style, arrange entries alphabetically by authors' last names. In the number style, arrange entries in numerical order—that is, in order of their citation in the text.

Format Begin the first line of each entry at the left margin and indent subsequent lines.

Authors List each author's name with the last name first, followed by initials for first and middle names. Do not use a comma between an author's last name and initials, and do not use periods or space with the initials. Do use a comma to separate authors' names.

Placement of dates In the name-year style, the date follows the author's or authors' names. In the number style, the date follows the publication information (for a book) or the periodical title (for a journal, magazine, or newspaper).

Journal titles Do not underline or italicize journal titles. For titles of two or more words, abbreviate words of six or more letters (without periods) and omit most prepositions, articles, and conjunctions. Capitalize each word. For example, *Journal of Chemical and Biochemical Studies* becomes J Chem Biochem Stud.

Book and article titles Do not underline, italicize, or use quotation marks around a book or an article title. Capitalize only the first word and any proper nouns. See model 2 on the next page.

Publication information for journal articles The name-year and number styles differ in the placement of the publication date (see above). However, both styles end with the journal's volume number, any issue number in parentheses, a colon, and the inclusive page numbers of the article, run together without space: 28:329-30 or 62(2):26-40.

The following examples show both a name-year reference and a number reference for each type of source.

CSE REFERENCE

Books

1. A book with one author

Gould SJ. 1987. Time's arrow, time's cycle. Cambridge: Harvard Univ Pr. 222 p.

1. Gould SJ. Time's arrow, time's cycle. Cambridge: Harvard Univ Pr; 1987. 222 p.

2. A book with two to ten authors

Hepburn PX, Tatin JM. 2002. Human physiology. New York: Columbia Univ Pr. 1026 p.

2. Hepburn PX, Tatin JM. Human physiology. New York: Columbia Univ Pr; 2002. 1026 p.

3. A book with more than ten authors

Evans RW, Bowditch L, Dana KL, Drummond A, Wildovitch WP, Young SL, Mills P, Mills RR, Livak SR, Lisi OL, and others. 1998. Organ transplants: ethical issues. Ann Arbor: Univ of Michigan Pr. 498 p.

3. Evans RW, Bowditch L, Dana KL, Drummond A, Wildovitch WP, Young SL, Mills P, Mills RR, Livak SR, Lisi OL, and others. Organ transplants: ethical issues. Ann Arbor: Univ of Michigan Pr; 1998. 498 p.

4. A book with an editor

Jonson P, editor. 2005. Anatomy yearbook. Los Angeles: Anatco. 628 p.

4. Jonson P, editor. Anatomy yearbook. Los Angeles: Anatco; 2005. 628 p.

5. A selection from a book

Krigel R, Laubenstein L, Muggia F. 2002. Kaposi's sarcoma. In: Ebbeson P, Biggar RS, Melbye M, editors. AIDS: a basic guide for clinicians. 2nd ed. Philadelphia: WB Saunders. p 100-26.

5. Krigel R, Laubenstein L, Muggia F. Kaposi's sarcoma. In: Ebbeson P, Biggar RS, Melbye M, editors. AIDS: a basic guide for clinicians. 2nd ed. Philadelphia: WB Saunders; 2002. p 100-26.

6. An anonymous work

[Anonymous]. 2004. Health care for multiple sclerosis. New York: US Health Care. 86 p.

6. [Anonymous]. Health care for multiple sclerosis. New York: US Health Care; 2004. 86 p.

7. Two or more cited works by the same author published in the same year

Gardner H. 1973a. The arts and human development. New York: J Wiley. 406 p.

Gardner H. 1973b. The quest for mind: Piaget, Lévi-Strauss, and the structuralist movement. New York: AA Knopf. 492 p.

(The number style does not require such forms.)

Periodicals: Journals, magazines, newspapers

8. An article in a journal with continuous pagination throughout the annual volume

Ancino R, Carter KV, Elwin DJ. 2002. Factors contributing

to viral immunity: a review of the research. Dev Biol 30:156-9.

8. Ancino R, Carter KV, Elwin DJ. Factors contributing to viral immunity: a review of the research. Dev Biol 2002;30:156-9.

9. An article in a journal that pages issues separately

Kim P. 1996 Feb. Medical decision making for the dying. Milbank Quar 64(2):26-40.

9. Kim P. Medical decision making for the dying. Milbank Quar 1996 Feb;64(2):26-40.

10. An article in a newspaper

Kolata G. 2001 Jan 7. Kill all the bacteria! New York Times;Sect B:1(col 3).

10. Kolata G. Kill all the bacteria! New York Times 2001 Jan 7;Sect B:1(col 3).

11. An article in a magazine

Scheiber N. 2002 June 24. Finger tip: why fingerprinting won't work. New Republic:15-6.

11. Scheiber N. Finger tip: why fingerprinting won't work. New Republic 2002 June 24:15-6.

Electronic sources

Scientific Style and Format includes a few formats for citing electronic sources, derived from *National Library of Medicine Recommended Formats for Bibliographic Citation.* For additional formats, the CSE Web site recommends the NLM 2001 supplement for Internet sources. The following models adapt these NLM formats to CSE name-year and number styles.

12. A source on CD-ROM or DVD-ROM

Reich WT, editor. 2005. Encyclopedia of bioethics [DVD]. New York: Co-Health.

12. Reich WT, editor. Encyclopedia of bioethics [DVD]. New York: Co-Health; 2005.

13. An online journal article

Grady GF. 2003. The here and now of hepatitis B immunization. Today's Med [Internet] [cited 2004 Dec 7];6(2):39-41. Available from: http://www.fmrt.org/todaysmedicine/Grady050293.pdf6

13. Grady GF. The here and now of hepatitis B immunization. Today's Med [Internet] 2003 [cited 2004 Dec 7];6(2): 39-41. Available from: http://www.fmrt.org/todaysmedicine/Grady050293.pdf6

Give the date of your access, preceded by "cited," in brackets: [cited 2004 Dec 7] in the models above.

14. An online book

Ruch BJ, Ruch DB. 2001. Homeopathy and medicine: resolving the conflict [Internet]. New York: Albert Einstein Coll of Medicine [cited 2005 Jan 28]. [about 50 p.]. Available from: http://www.einstein.edu/medicine/books/ruch.html

14. Ruch BJ, Ruch DB. Homeopathy and medicine: resolving the conflict [Internet]. New York: Albert Einstein Coll of Medicine; 2001 [cited 2005 Jan 28]. [about 50 p.]. Available from: http:// www.einstein.edu/medicine/books/ruch.html

15. A source retrieved from an online database

McAskill MR, Anderson TJ, Jones RD. 2002. Saccadic adaptation in neurological disorders. Prog Brain Res 140:417-31. In: PubMed [Internet]. Bethesda (MD): National Library of Medicine; [cited 2004 Mar 6]. Available from: http://www.ncbi.nlm.nih.gov/PubMed; PMID: 12508606.

15. McAskill MR, Anderson TJ, Jones RD. Saccadic adaptation in neurological disorders. Prog Brain Res 2002;140:417-31. In: PubMed [Internet]. Bethesda (MD): National Library of Medicine; [cited 2004 Mar 6]. Available from: http://www.ncbi.nlm.nih.gov/PubMed; PMID: 12508606.

16. A Web site

American Medical Association [Internet]. 2005. Chicago: American Medical Association; [cited 2005 Jan 2]. Available from: http://ama-assn.org

16. American Medical Association [Internet]. Chicago: American Medical Association; 2005 [cited 2005 Jan 2]. Available from: http://ama-assn.org

17. Electronic mail

Millon M. 2004 May 4. Grief therapy [Internet]. Message to: Naomi Sakai. 3:16 pm [cited 2004 May 4]. [about 2 screens].

17. Millon M. Grief therapy [Internet]. Message to: Naomi

Sakai. 2004 May 4, 3:16 pm [cited 2004 May 4]. [about 2 screens].

18. A posting to a discussion list

Stalinsky Q. 2002 Aug 16. The hormone-replacement study. In: Women Physicians Congress [Internet]. [Chicago: American Medical Association]; 9:26 am [cited 2004 Aug 17]. [about 8 paragraphs]. Available from: ama-wpc@ama-assn.org

18. Stalinsky Q. The hormone-replacement study. In: Women Physicians Congress [Internet]. [Chicago: American Medical Association]; 2004 Aug 16, 9:26 am [cited 2004 Aug 17]. [about 8 paragraphs]. Available from: ama-wpc@ama-assn.org

Other sources

19. A government publication

Committee on Science and Technology, House (US). 1999. Hearing on procurement and allocation of human organs for transplantation. 106th Cong., 1st Sess. House Doc. nr 409.

19. Committee on Science and Technology, House (US). Hearing on procurement and allocation of human organs for transplantation. 106th Cong., 1st Sess. House Doc. nr 409; 1999.

20. A nongovernment report

Warnock M. 2001. Report of the Committee on Fertilization and Embryology. Baylor University, Department of Embryology. Waco (TX): Baylor Univ. Report nr BU/DE.4261.

20. Warnock M. Report of the Committee on Fertilization and Embryology. Baylor University, Department of Embryology. Waco (TX): Baylor Univ; 2001. Report nr BU/DE.4261.

21. A sound recording, video recording, or film

Teaching Media. 2000. Cell mitosis [videocassette]. White Plains (NY): Teaching Media. 1 videocassette: 40 min, sound, black and white, 1/2 in.

21. Cell mitosis [videocassette]. White Plains (NY): Teaching Media; 2000. 1 videocassette: 40 min, sound, black and white, 1/2 in.

PART VI

Special Writing Situations

40 Designing Documents

Legible, consistent, and attractive papers and correspondence serve your readers and reflect well on you. This chapter shows the basics of formatting any document clearly and effectively.

40a ▪ Formats for academic papers

Many academic disciplines prefer specific formats for students' papers. This book details three such formats:

- MLA, used in English, foreign languages, and some other humanities (pp. 159–62).
- APA, used in the social sciences and some natural sciences (pp. 175–78).
- Chicago, used in history, art history, religion, and some other humanities (pp. 189–91).

The design guidelines in this chapter extend the range of elements and options covered by most academic styles. Your instructors may want you to adhere strictly to a particular style or may allow some latitude in design. Ask them for their preferences.

40b ▪ Clear and effective documents

Your papers, reports, and correspondence must of course be neat and legible. But you can do more to make your work accessible and attractive by taking care with margins, text, lists, headings, and illustrations (tables, figures, images).

Margins

Provide minimum one-inch margins on all sides of a page to prevent unpleasant crowding. If your document will be presented in a binder, provide a larger left margin—say, 1½ inches.

Text

Line spacing

Double-space most academic documents, with an initial indention for paragraphs. Single-space most business

Visit *www.ablongman.com/littlebrown* for added help with document design.

documents, with an extra line of space between paragraphs. Double- or triple-space to set off headings in both kinds of writing.

Type fonts and sizes

The readability of text also derives from the type fonts (or faces) and their sizes. For academic and business documents, choose a standard font and a type size of 10 or 12 points, as in the samples below.

10-point Courier	10-point Times Roman
12-point Courier	12-point Times Roman

Highlighting

Within a document's text underlined, *italic,* **boldface,** or even color type can emphasize key words or sentences. Underlining is rarest these days, having been replaced by italics in business and most academic writing. (It remains called for in MLA style. See Chapter 36.) In both academic and business writing, boldface can give strong emphasis—for instance, to a term being defined. More in business than in academic writing, color can highlight headings and illustrations.

Lists

Lists give visual reinforcement to the relations between like items—for example, the steps in a process or the elements of a proposal. A list is easier to read than a paragraph and adds white space to the page.

When wording a list, work for parallelism among items—for instance, all complete sentences or all phrases (see also p. 19). Set the list with space above and below and with numbering or bullets (centered dots or other devices, such as the small squares in the list below and on the next page). Most word processors can format a numbered or bulleted list automatically.

Headings

Headings are signposts. In a long or complex document, they direct the reader's attention by focusing the eye on a document's most significant content. Most academic and business documents use headings functionally, to divide text, orient readers, and create emphasis. Follow these guidelines:

- *Use one, two, or three levels of headings* depending on the needs of your material and the length of your document. Some level of heading every two or so pages will help keep readers on track.

- *Create an outline of your document* to plan where headings should go. Reserve the first level of heading for the main points (and sections) of your document. Use a second and perhaps a third level of heading to mark subsections of supporting information.
- *Keep headings as short as possible* while making them specific about the material that follows.
- *Word headings consistently*—for instance, all questions (*What Is the Scientific Method?*), all phrases° with *-ing* words (*Understanding the Scientific Method*), or all phrases with nouns (*The Scientific Method*).
- *Indicate the relative importance of headings* with type size, positioning, and highlighting, such as capital letters, underlining, or boldface.

FIRST-LEVEL HEADING

Second-Level Heading

Third-Level Heading

- *Keep the appearance simple,* minimizing variations in type style and type size and avoiding extra-large letters or unusual styles.
- *Don't break a page immediately after a heading*. Push the heading to the next page.

Note Document format in psychology and some other social sciences requires a particular treatment of headings. See page 176.

Tables, figures, and images

Tables, figures, and images can often make a point for you more efficiently and effectively than words can. Tables present data. Figures (such as graphs or charts) usually recast data in visual form. Images (such as diagrams, drawings, photographs, and clip art) can explain processes, represent what something looks like, add emphasis, or convey a theme.

Follow these guidelines when using tables, figures, or images in academic and most business writing:

- *Focus on a purpose for your table or illustration*—a reason for including it and a point you want it to make. Otherwise, readers may find it irrelevant or confusing.
- *Provide a source note for someone else's independent material,* whether it's data or a complete illustration (see p. 138). Each discipline has a slightly different style for such source notes; those in the table opposite and the figures on page 204 reflect the style of the social sciences. See also Chapters 36–39.

- *Number figures and images together,* and label them as figures: Figure 1, Figure 2, and so on.
- *Number and label tables separately from figures*: Table 1, Table 2, and so on.
- *Refer to each illustration* (for instance, "See Figure 2") at the point(s) in the text where readers will benefit by consulting it.
- *Place each illustration on a page by itself* immediately after the page that refers to it.

Many organizations and academic disciplines have preferred styles for tables and figures that differ from those given here. When in doubt about how to prepare tables and figures, ask your instructor or supervisor.

Tables

Tables usually summarize raw data, displaying the data concisely and clearly.

- *Provide a self-explanatory title above the table.*
- *Provide self-explanatory headings for horizontal rows and vertical columns.* Use abbreviations only if you are certain readers will understand them.
- *Lay out rows and columns for maximum clarity.* In the sample below, for instance, lines divide the table into parts, headings align with their data, and numbers align vertically down columns.

Table 1

Percentage of Young Adults Living at Home, 1960–2000

	1960	1970	1980	1990	2000
Males					
Age 18–24	52	54	54	58	57
Age 25–34	9	9	10	15	13
Females					
Age 18–24	35	41	43	48	47
Age 25–34	7	7	7	8	8

Note. Data from U.S. Bureau of Census, *Census 2000 Summary Tables*, retrieved July 1, 2004, from http://www.census.gov/servlet/QTTTable?_ts=30543101060

Figures

Figures represent data graphically. They include the three kinds presented on the following page: pie charts (showing percentages making up a whole), bar graphs (showing comparative data), and line graphs (showing change).

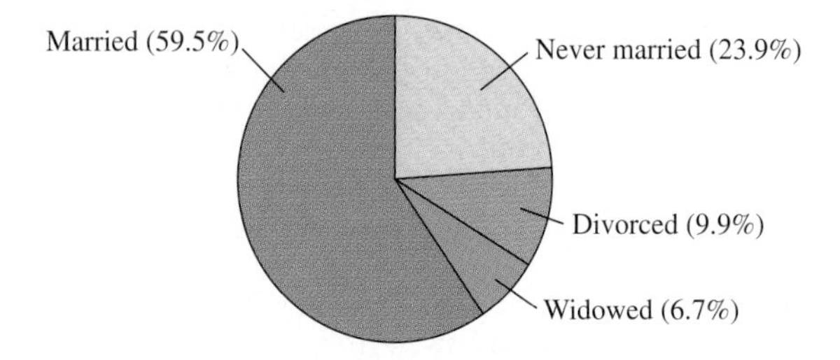

Figure 1. Marital status in 2003 of US adults aged eighteen and over. Data from *Statistical Abstract of the United States, 2004.*

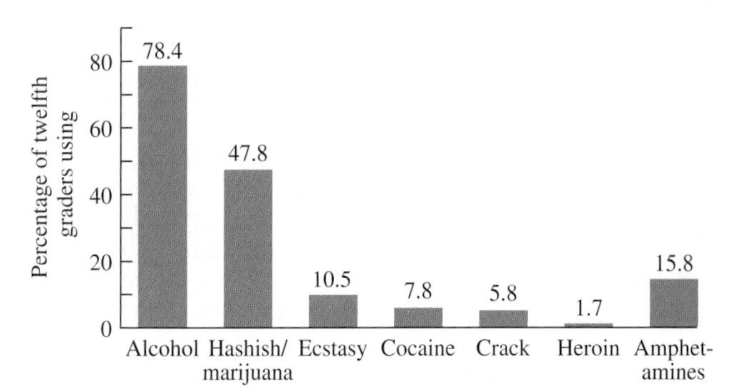

Figure 2. Lifetime prevalence of use of alcohol, compared with other drugs, among twelfth graders in 2003. Data from Monitoring the Future, University of Michigan, *Monitoring the Future: A Continuing Study of American Youth,* retrieved October 10, 2004, from http://www.monitoringthefuture.org/03data.html#2003data-drugs

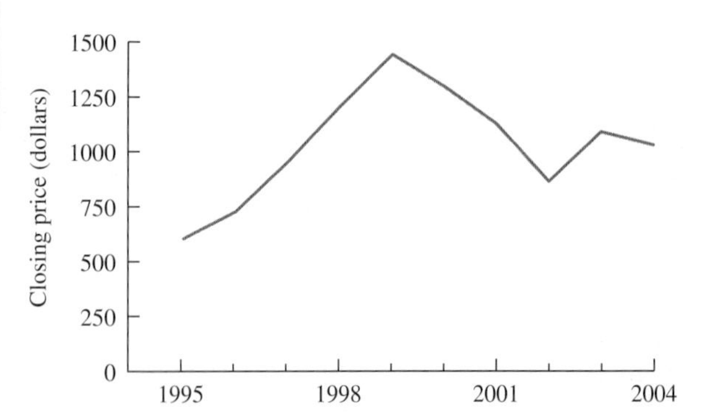

Figure 3. Standard & Poor's 500 Index year-end prices, 1995–2004. Data from *S&P Indices,* retrieved January 26, 2005, from Standard & Poor's Web site: http://www2.standardandpoors.com/NASApp/cs/Index=500

- *Provide a self-explanatory caption or legend below the figure.*
- *Provide self-explanatory labels for all parts of the figure.*
- *Draw the figure to reflect its purpose and the visual effect you want it to have.* In Figure 3 on the facing page, shortening the horizontal date axis emphasizes the movement of the line over time.

Photographs and other images

Diagrams, photographs, and other images can add substance to an academic or business document. In a psychology paper, for instance, a photograph may illustrate a key experiment. But images cannot represent your ideas by themselves: you need to relate them to your text, explain their significance, and label, number, and caption them (see pp. 202–03).

Note When using an image prepared by someone else, you must always cite the source. Most copyright holders allow the use of their material for academic papers that are distributed only within a class. However, public use, such as on a Web site, requires permission from the copyright holder.

41 Writing Online

The main forms of online writing—electronic mail and Web compositions—dramatically expand your options as a writer. This chapter addresses some special concerns of writing online.

41a ▪ Electronic mail

The guidelines below apply to various e-mail functions, from conversing with friends to conducting serious research.

Sending messages

- *Give your message a subject heading that describes its content.* Then your reader knows what priority to assign the message—a particular help for readers who receive many messages a day.

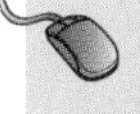

Visit *www.ablongman.com/littlebrown* for added help with writing online.

- *Compose a message that is concise and relevant to your recipients' concerns*. Adjust the formality of your message to your writing situation. In the message below, the writer knows the recipients well and yet has serious information to convey to them, so he writes informally and yet states his points carefully.

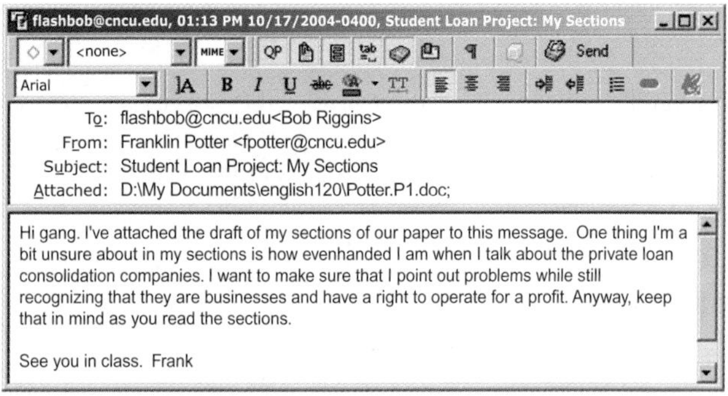

flashbob@cncu.edu, 01:13 PM 10/17/2004-0400, Student Loan Project: My Sections

<none> MIME QP Send

Arial

To: flashbob@cncu.edu<Bob Riggins>
From: Franklin Potter <fpotter@cncu.edu>
Subject: Student Loan Project: My Sections
Attached: D:\My Documents\english120\Potter.P1.doc;

Hi gang. I've attached the draft of my sections of our paper to this message. One thing I'm a bit unsure about in my sections is how evenhanded I am when I talk about the private loan consolidation companies. I want to make sure that I point out problems while still recognizing that they are businesses and have a right to operate for a profit. Anyway, keep that in mind as you read the sections.

See you in class. Frank

- *Use short paragraphs with blank lines between them*. For long messages—which recipients can review only one screen at a time—a tight structure, a clear forecast of the content, and a clear division into parts (using headings if necessary) not only improve effectiveness but also show courtesy.
- *Proofread* all but the most informal messages for errors in grammar, punctuation, and spelling.
- *Be aware that file attachments are not always readable*. If you have trouble with attachments, consult the technology advisers at your school.

Note Some e-mail programs don't allow underlining, italics, and boldface. Even if you can use such highlighting, you should assume that your recipients will not be able to see it in your messages. For alternatives, use underscores to indicate _underlining_ or asterisks to provide *emphasis*.

Responding to messages

- *Check that your reply has an appropriate subject heading*. Most e-mail programs label a response with the same subject heading as the original, preceded by *Re:* (from Latin, meaning "in reference to"). If your response indeed continues the same subject, then *Re:* indicates as much. However, if you raise a new issue, you should rewrite the subject heading to say so.
- *Use quoted material from earlier messages critically*. By weaving your replies into the material quoted from someone else's message, you can respond to the

author point by point, as you would in conversation. However, delete from the original anything you are not responding to so that your recipient can focus on what you have to say without wading through his or her own words.

Observing netiquette

You won't always see others observing **netiquette**, or Internet etiquette, but you will see that those who do observe it receive the more thoughtful and considerate replies.

- *Pay careful attention to tone*. Refrain from **flaming**, or attacking, correspondents. Don't use all-capital letters, which SHOUT. And use irony or sarcasm only cautiously: in the absence of facial expressions, they can lead to misunderstandings.
- *Avoid participating in flame "wars,"* overheated dialogs that contribute little or no information or understanding. If a war breaks out in a discussion, ignore it: don't rush to defend someone who is being attacked and don't respond even if you are under attack yourself.
- *Be a forgiving reader*. Avoid nitpicking over spelling or other surface errors. And because attitudes are sometimes difficult to convey, give authors an initial benefit of the doubt: a writer who at first seems hostile may simply have tried too hard to be concise; a writer who at first seems unserious may simply have failed at injecting humor into a worthwhile message.
- *Forward a message only with the permission of the author.*
- *Avoid spamming*. With a few keystrokes, you can broadcast a message to many recipients at once. Occasionally you may have important information that everyone on a list will want to know. But flooding whole lists with irrelevant messages—called **spamming**—is rude and irritating.

41b ▪ Web composition

Unlike traditional documents, which are meant to be read in sequence from start to finish, most Web sites are intended to be examined in whatever order readers choose as they follow links within the site and to other sites. A Web site thus requires careful planning of the links and thoughtful cues to orient readers:

- *Sketch possible site plans before getting started*, using a branching diagram to show major divisions and connections.

- *Plan a menu for each of the site's pages* that overviews the organization and links directly to other pages.
- *Distill your text* to include only essential information, and use headings to break the text into accessible chunks. Readers have difficulty following long, uninterrupted text passages on screen.
- *Standardize elements of the design* so that readers know what to expect and can scan your pages easily.
- *Use icons, photographs, artwork, and other visual elements for a purpose, not merely for decoration.* Use captions to relate the visuals to your text.

Most Web pages are created using hypertext markup language, or HTML, and an HTML editor that works much as a word processor does. For information about HTML editors and added help with creating Web sites, visit this book's Web site at *www.ablongman.com/littlebrown.*

42 Making Oral Presentations

Effective speakers use organization, voice, body language, and other techniques to help their audiences listen to their presentations.

42a ▪ Organization

Give your oral presentation a recognizable shape so that listeners can see how ideas and details relate to each other.

The introduction

The beginning of an oral presentation should try to accomplish three goals:

- *Gain the audience's attention and interest.* Begin with a question, an unusual example or statistic, or a short, relevant story.
- *Put yourself in the speech.* Demonstrate your expertise, experience, or concern to gain the interest and trust of your audience.

Visit ***www.ablongman.com/littlebrown*** for added help with oral presentations and *PowerPoint.*

- *Introduce and preview your topic and purpose.* By the time your introduction is over, listeners should know what your subject is and the direction you'll take to develop your ideas.

Your introduction should prepare your audience for your main points but not give them away. Think of it as a sneak preview of your speech, not the place for an apology such as *I wish I'd had more time to prepare* . . . or a dull statement such as *My speech is about*. . . .

Supporting material

Just as you do when writing, you should use facts, statistics, examples, and expert opinions to support the main points of your oral presentation. In addition, you can make your points more memorable with vivid description, well-chosen quotations, true or fictional stories, and analogies.

The conclusion

You want your conclusion to be clear, of course, but you also want it to be memorable. Remind listeners of how your topic and main idea connect to their needs and interests. If your speech was motivational, tap an emotion that matches your message. If your speech was informational, give some tips on how to remember important details.

42b ▪ Delivery

Methods of delivery

You can deliver an oral presentation in several ways:

- *Impromptu, without preparation:* Make a presentation without planning what you will say. Impromptu speaking requires confidence and excellent general preparation.
- *Extemporaneously:* Prepare notes to glance at but not read from. This method allows you to look and sound natural while ensuring that you don't forget anything.
- *Speaking from a text:* Read aloud from a written presentation. You won't lose your way, but you may lose your audience. Avoid reading for an entire presentation.
- *Speaking from memory:* Deliver a prepared presentation without notes. You can look at your audience every minute, but the stress of retrieving the next words may make you seem tense and unresponsive.

Vocal delivery

The sound of your voice will influence how listeners receive you. Rehearse your presentation several times until you are confident that you are speaking loudly, slowly, and clearly enough for your audience to understand you.

Physical delivery

You are more than your spoken words when you make an oral presentation. If you are able, stand up to deliver your presentation, moving your body toward one side of the room and the other, stepping out from behind any lectern or desk, and gesturing as appropriate. Above all, make eye contact with your audience as you speak. Looking directly in your listeners' eyes conveys your honesty, your confidence, and your control of the material.

Visual aids

You can supplement an oral presentation with visual aids such as posters, models, slides, or videos.

- *Use visual aids to underscore your points.* Short lists of key ideas, illustrations such as graphs or photographs, or objects such as models can make your presentation more interesting and memorable. But use visual aids judiciously: a battery of illustrations or objects will bury your message rather than amplify it.
- *Coordinate visual aids with your message.* Time each visual to reinforce a point you're making. Tell listeners what they're looking at. Give them enough viewing time so they don't mind turning their attention back to you.
- *Show visual aids only while they're needed.* To regain your audience's attention, remove or turn off any aid as soon as you have finished with it.

With *PowerPoint* software, you can integrate visual aids with your speaking notes distilled as pointers for the audience. Learn about using *PowerPoint* at this book's Web site: *www.ablongman.com/littlebrown.*

Practice

Take time to rehearse your presentation out loud, with the notes you will be using. Gauge your performance by making an audio- or videotape of yourself or by practicing in front of a mirror. Practicing out loud will also tell you if your presentation is running too long or too short.

If you plan to use visual aids, you'll need to practice with them, too. Your goal is to eliminate hitches (upside-down slides, missing charts) and to weave the visuals seamlessly into your presentation.

Stage fright

Many people report that speaking in front of an audience is their number-one fear. Even many experienced and polished speakers have some anxiety about delivering an oral presentation, but they use this nervous energy to their advantage, letting it propel them into working hard on each presentation. Several techniques can help you reduce anxiety:

- *Use simple relaxation exercises.* Deep breathing or tensing and relaxing your stomach muscles can ease some of the physical symptoms of speech anxiety—stomachache, rapid heartbeat, and shaky hands, legs, and voice.
- *Think positively.* Instead of worrying about the mistakes you might make, concentrate on how well you've prepared and practiced your presentation and how significant your ideas are.
- *Don't avoid opportunities to speak in public.* Practice and experience build speaking skills and offer the best insurance for success.

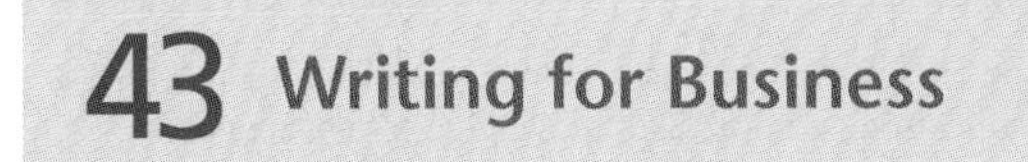

43 Writing for Business

The principles of document design covered in Chapter 40 apply to most business writing. This chapter discusses business letters, résumés, and business memos. For help with electronic mail, see pages 205–07.

43a ▪ Job-application letter

In any letter to a businessperson, you are addressing someone who wants to see quickly why you are writing and how to respond to you. For a job application, see the sample on the next page and use the following guidelines.

Visit *www.ablongman.com/littlebrown* for added help with business writing.

Minimum 1"

Return-address heading

3712 Swiss Avenue
Dallas, TX 75204
March 4, 2005

Double-space

Raymond Chipault
Human Resources Manager
Dallas News
Communications Center
Dallas, TX 75222

Inside address

Double-space

Dear Mr. Chipault: — Salutation

Double-space

In response to your posting in the English Department of Southern Methodist University, I am applying for the summer job of part-time editorial assistant for the *Dallas News*.

Double-space

I am now enrolled at Southern Methodist University as a sophomore, with a dual major in English literature and journalism. My courses so far have included news reporting, copy editing, and electronic publishing. I worked a summer as a copy aide for my hometown newspaper, and for two years I have edited and written sports stories and features for the university newspaper. My feature articles cover subjects as diverse as campus elections, parking regulations, visiting professors, and speech codes.

1" 1"

Double-space

As the enclosed résumé and writing samples indicate, my education and practical knowledge of newspaper work prepare me for the opening you have.

Double-space

I am available for an interview at your convenience and would be happy to show more samples of my writing. Please call me at 214-744-3816 or e-mail me at ianirv@mail.smu.edu.

Double-space

Sincerely, — Close

Quadruple-space

Ian M. Irvine — Signature

Ian M. Irvine

Enc.

Content

- *Interpret your résumé for the particular job*. Instead of reciting your job history, highlight and reshape only the relevant parts.
- *Announce at the outset what job you seek and how you heard about it.*
- *Include any special reason you have for applying,* such as a specific career goal.
- *Summarize your qualifications for the job,* including relevant facts about education and employment and emphasizing notable accomplishments. Mention that additional information appears in an accompanying résumé.
- *Describe your availability*. At the end of the letter, mention that you are available for an interview at the convenience of the addressee, or specify when you will be available.

Format

- *Use standard paper,* either unlined white paper measuring 8½ inches by 11 inches or letterhead stationery with your address printed at the top of the sheet.
- *Type the letter single-spaced*. Use double space between elements and paragraphs. Use only one side of a sheet.
- *Address your letter to a specific person*. Call the company or department to ask whom to address. If you can't find a person's name, then use a job title (*Dear Human Resources Manager*) or a general salutation (*Dear Smythe Shoes*).
- *Close the letter with an expression that reflects the formality of the situation*. *Respectfully, Cordially,* and *Sincerely* are more formal than *Best wishes* or *Regards*.
- *Use an envelope that will accommodate the letter once it is folded horizontally in thirds*. Show your name and address in the upper-left corner and the addressee's name, title, and address in the center.

43b ▪ Résumé

For the résumé that accompanies your letter of application, you can use the sample on the next page and the following guidelines.

- *Provide the appropriate information in table form:* your name and address, career objective, education, employment history, any special skills or awards, and information about how to obtain your references.

Ian M. Irvine
3712 Swiss Avenue
Dallas, TX 75204
214-744-3816
ianirv@mail.smu.edu

Position desired
Part-time editorial assistant.

Education
Southern Methodist University, 2003 to present.
Current standing: sophomore.
Major: English literature and journalism.
Journalism courses: news reporting, copy editing, electronic publishing, communications arts, broadcast journalism.

Abilene (Texas) Senior High School, 1999-2003.
Graduated with academic, college-preparatory degree.

Employment history
2003 to present. Reporter, *Daily Campus*, student newspaper of Southern Methodist University.
Write regular coverage of baseball, track, and soccer teams. Write feature stories on campus policies and events. Edit sports news, campus listings, features.

Summer 2004. Copy aide, *Abilene Reporter-News*.
Routed copy, ran errands, and assisted reporters with research.

Summer 2003. Painter, Longhorn Painters, Abilene.
Prepared and painted exteriors and interiors of houses.

Special skills
Fluent in Spanish.
Proficient in Internet research and word processing.

References
Available upon request:

Placement Office
Southern Methodist University
Dallas, TX 75275

- *Use headings to mark the sections of the résumé.* Space around the headings and within sections so that important information stands out.
- *Use capital letters conventionally.* Passages with many capitals can be hard to read. Do use capitals for proper nouns (pp. 95–96), but drop them for job titles, course names, department names, and the like.
- *Limit your résumé to one page so that it can be quickly reviewed.* However, if your experience and education are extensive, a two-page résumé is preferable to a single cramped, unreadable page.

Employers often want an electronic copy of a résumé so that they can add it to a computerized database of applicants. They may scan a paper copy to convert it to an electronic file, or they may request electronic copy from you in the first place. If you think a potential employer may use an electronic version of your résumé, follow these additional guidelines:

- *Keep the design simple* for accurate scanning or electronic transmittal. Avoid images, unusual type, more than one column, vertical or horizontal lines, and highlighting (boldface, italic, or underlining). If its highlighting were removed, the traditionally designed sample opposite could probably be scanned or transmitted electronically.
- *Use concise, specific words to describe your skills and experience.* The employer's computer may use keywords (often nouns) to identify the résumés of suitable job candidates, and you want to ensure that your résumé includes the appropriate keywords. Name your specific skills—for example, the computer programs you can operate—and write concretely with words like *manager* (not *person with responsibility for*) and *reporter* (not *staff member who reports*). Look for likely keywords in the employer's description of the job you seek.

43c ▪ Business memo

Business memos (short for memorandums) address people within the same organization. A memo can be quite long, but more often it deals briefly with a specific topic, such as an answer to a question, a progress report, or an evaluation. Both the content and the format of a memo aim to get to the point and dispose of it quickly, as in the sample on the next page.

Bigelow Wax Company

TO: Aileen Rosen, Director of Sales
FROM: Patricia Phillips, Territory 12 PP
DATE: February 12, 2005
SUBJECT: 2004 sales of Quick Wax in Territory 12

Since it was introduced in January of 2004, Quick Wax has been unsuccessful in Territory 12 and has not affected the sales of our Easy Shine. Discussions with customers and my own analysis of Quick Wax suggest three reasons for its failure to compete with our product.

1. Quick Wax has not received the promotion necessary for a new product. Advertising—primarily on radio—has been sporadic and has not developed a clear, consistent image for the product. In addition, the Quick Wax sales representative in Territory 12 is new and inexperienced; he is not known to customers, and his sales pitch (which I once overheard) is weak. As far as I can tell, his efforts are not supported by phone calls or mailings from his home office.

2. When Quick Wax does make it to the store shelves, buyers do not choose it over our product. Though priced competitively with our product, Quick Wax is poorly packaged. The container seems smaller than ours, though in fact it holds the same eight ounces. The lettering on the Quick Wax package (red on blue) is difficult to read, in contrast to the white-on-green lettering on the Easy Shine package.

3. Our special purchase offers and my increased efforts to serve existing customers have had the intended effect of keeping customers satisfied with our product and reducing their inclination to stock something new.

Copies: L. Mendes, Director of Marketing
L. MacGregor, Customer Service Manager

- *State your reason for writing in the first sentence.*
- *Make your first paragraph work.* Concisely present your solution, recommendation, answer, or evaluation.
- *Deliver the support in the body of the memo.* The paragraphs may be numbered or bulleted so that the main divisions of your message are easy to see. In a long memo, you may need headings (see pp. 201–02).
- *Suit your style and tone to your audience.* For instance, you'll want to address your boss or a large group of readers more formally than you would a coworker who is also a friend.
- *Use a conventional memo heading.* Include the company's name, the addressee's name, the writer's name (initialed in handwriting), the date, and a subject description or title.
- *Type the body of the memo single-spaced.* Double-space between paragraphs, with no paragraph indentions.
- *List the people receiving copies of the memo* two spaces below the last line.

Glossary of Usage

This glossary provides notes on words or phrases that often cause problems for writers. The recommendations for standard written English are based on current dictionaries and usage guides. Items labeled *nonstandard* should be avoided in final drafts of academic and business writing. Those labeled *colloquial* and *slang* appear in some informal writing and may occasionally be used for effect in more formal academic and career writing. (Words and phrases labeled *colloquial* include those labeled *informal* by many dictionaries.) See Chapter 5 for more on levels of language.

a, an Use *a* before words beginning with consonant sounds: *a historian, a one-o'clock class, a university.* Use *an* before words that begin with vowel sounds, including silent *h*'s: *an orgy, an L, an honor.*

The article before an abbreviation depends on how the abbreviation is read: *She was once an AEC aide* (*AEC* is read as three separate letters); *Many Americans opposed a SALT treaty* (*SALT* is read as one word, *salt*).

See also pp. 56–58 on the uses of *a/an* versus *the.*

accept, except *Accept* is a verb° meaning "receive." *Except* is usually a preposition° or conjunction° meaning "but for" or "other than"; when it is used as a verb, it means "leave out." *I can accept all your suggestions except the last one. I'm sorry you excepted my last suggestion from your list.*

advice, advise *Advice* is a noun,° and *advise* is a verb.° *Take my advice; do as I advise you.*

affect, effect Usually *affect* is a verb,° meaning "to influence," and *effect* is a noun, meaning "result": *The drug did not affect his driving; in fact, it seemed to have no effect at all.* (Note that *effect* occasionally is used as a verb meaning "to bring about": *Her efforts effected a change.* And *affect* is used in psychology as a noun meaning "feeling or emotion": *One can infer much about affect from behavior.*)

all, always, never, no one These absolute words often exaggerate a situation in which *many, often, rarely,* or *few* is more accurate.

all ready, already *All ready* means "completely prepared," and *already* means "by now" or "before now": *We were all ready to go to the movie, but it had already started.*

all right *All right* is always two words. *Alright* is a common misspelling.

all together, altogether *All together* means "in unison," or "gathered in one place." *Altogether* means "entirely." *It's not altogether true that our family never spends vacations all together.*

allusion, illusion An *allusion* is an indirect reference, and an *illusion* is a deceptive appearance: *Paul's constant allusions to Shakespeare created the illusion that he was an intellectual.*

a lot *A lot* is always two words, used informally to mean "many." *Alot* is a common misspelling.

always See *all, always, never, no one.*

among, between In general, use *between* only for relationships of two and *among* for more than two.

amount, number Use *amount* with a singular noun that names something not countable (a noncount noun°): *The amount of food varies.* Use *number* with a plural noun that names more than one of something countable (a count noun°): *The number of calories must stay the same.*

and/or *And/or* indicates three options: one or the other or both (*The decision is made by the mayor and/or the council*). If you mean all three options, *and/or* is appropriate. Otherwise, use *and* if you mean both, *or* if you mean either.

anxious, eager *Anxious* means "nervous" or "worried" and is usually followed by *about. Eager* means "looking forward" and is usually followed by *to. I've been anxious about getting blisters. I'm eager* [not *anxious*] *to get new cross-training shoes.*

anybody, any body; anyone, any one *Anybody* and *anyone* are indefinite pronouns;° *any body* is a noun° modified by *any; any one* is a pronoun° or adjective° modified by *any. How can anybody communicate with any body of government? Can anyone help Amy? She has more work than any one person can handle.*

any more, anymore *Any more* means "no more"; *anymore* means "now." Both are used in negative constructions: *He doesn't want any more. She doesn't live here anymore.*

anyways, anywheres Nonstandard for *anyway* and *anywhere.*

are, is Use *are* with a plural subject° (*books are*), *is* with a singular subject (*book is*). See pp. 43–46.

as Substituting for *because, since,* or *while, as* may be vague or ambiguous: *As we were stopping to rest, we decided to eat lunch.* (Does *as* mean "while" or "because"?) *As* should never be used as a substitute for *whether* or *who. I'm not sure whether* [not *as*] *we can make it. That's the man who* [not *as*] *gave me directions.*

as, like See *like, as.*

at this point in time Wordy for *now, at this point,* or *at this time.*

awful, awfully Strictly speaking, *awful* means "inspiring awe." As intensifiers meaning "very" or "extremely" (*He tried awfully hard*), *awful* and *awfully* should be avoided in formal speech or writing.

a while, awhile *Awhile* is an adverb;° *a while* is an article° and a noun.° *I will be gone awhile* [not *a while*]. *I will be gone for a while* [not *awhile*].

bad, badly In formal speech and writing, *bad* should be used only as an adjective;° the adverb° is *badly. He felt bad because his tooth ached badly.* In *He felt bad,* the verb *felt* is a linking verb° and the adjective *bad* modifies the subject° *he,* not the verb *felt.* See also p. 54.

being as, being that Colloquial for *because,* the preferable word in formal speech or writing: *Because* [not *Being as*] *the world is round, Columbus never did fall off the edge.*

beside, besides *Beside* is a preposition° meaning "next to." *Besides* is a preposition meaning "except" or "in addition to" as well as an adverb° meaning "in addition." *Besides, several other people besides you want to sit beside Dr. Christensen.*

between, among See *among, between.*

bring, take Use *bring* only for movement from a farther place to a nearer one and *take* for any other movement. *First, take these books to the library for renewal, then take them to Mr. Daniels. Bring them back to me when he's finished.*

can, may Strictly, *can* indicates capacity or ability, and *may* indicates permission: *If I may talk with you a moment, I believe I can solve your problem.*

climatic, climactic *Climatic* comes from *climate* and refers to weather: *Last winter's temperatures may indicate a climatic change. Climactic* comes from *climax* and refers to a dramatic high point: *During the climactic duel between Hamlet and Laertes, Gertrude drinks poisoned wine.*

complement, compliment To *complement* something is to add to, complete, or reinforce it: *Her yellow blouse complemented her black hair.* To *compliment* something is to make a flattering remark about it: *He complimented her on her hair. Complimentary* can also mean "free": *complimentary tickets.*

conscience, conscious *Conscience* is a noun° meaning "a sense of right and wrong"; *conscious* is an adjective° meaning "aware" or "awake." *Though I was barely conscious, my conscience nagged me.*

continual, continuous *Continual* means "constantly recurring": *Most movies on television are continually interrupted by commercials. Continuous* means "unceasing": *Some cable channels present movies continuously without commercials.*

could of See *have, of.*

criteria The plural of *criterion* (meaning "standard for judgment"): *Our criteria are strict. The most important criterion is a sense of humor.*

data The plural of *datum* (meaning "fact"). Though *data* is often used with a singular verb, many readers prefer the plural verb and it is always correct: *The data fail* [not *fails*] *to support the hypothesis.*

device, devise *Device* is the noun,° and *devise* is the verb:° *Can you devise some device for getting his attention?*

different from, different than *Different from* is preferred: *His purpose is different from mine.* But *different than* is widely accepted when a construction using *from* would be wordy: *I'm a different person now than I used to be* is preferable to *I'm a different person now from the person I used to be.*

disinterested, uninterested *Disinterested* means "impartial": *We chose Pete, as a disinterested third party, to decide who was right. Uninterested* means "bored" or "lacking interest": *Unfortunately, Pete was completely uninterested in the question.*

don't *Don't* is the contraction for *do not,* not for *does not: I don't care, you don't care,* and *he doesn't* [not *don't*] *care.*

due to *Due* is an adjective° or noun;° thus *due to* is always acceptable as a subject complement:° *His gray hairs were due to age.* Many object to *due to* as a preposition° meaning "because of" (*Due to the holiday, class was canceled*). A rule of thumb is that *due to* is always correct after a form of the verb *be* but questionable otherwise.

eager, anxious See *anxious, eager.*

effect See *affect, effect.*

elicit, illicit *Elicit* is a verb° meaning "bring out" or "call forth." *Illicit* is an adjective° meaning "unlawful." *The crime elicited an outcry against illicit drugs.*

emigrate, immigrate *Emigrate* means "to leave one place and move to another": *The Chus emigrated from Korea. Immigrate* means "to move into a place where one was not born": *They immigrated to the United States.*

enthused Sometimes used colloquially as an adjective° meaning "showing enthusiasm." The preferred adjective is *enthusiastic: The coach was enthusiastic* [not *enthused*] *about the team's victory.*

etc. *Etc.,* the Latin abbreviation for "and other things," should be avoided in formal writing and should not be used to refer to people. When used, it should not substitute for precision, as in *The government provides health care, etc.,* and it should not end a list beginning *such as* or *for example.*

everybody, every body; everyone, every one *Everybody* and *everyone* are indefinite pronouns:° *Everybody* [or *Everyone*] *knows Tom steals. Every one* is a pronoun° modified by *every,* and *every body* a noun° modified by *every.* Both refer to each thing or person of a specific group and are typically followed by *of: The game commissioner has stocked every body of fresh water in the state with fish, and now every one of our rivers is a potential trout stream.*

everyday, every day *Everyday* is an adjective° meaning "used daily" or "common"; *every day* is a noun° modified by *every: Everyday problems tend to arise every day.*

everywheres Nonstandard for *everywhere.*

except See *accept, except.*

explicit, implicit *Explicit* means "stated outright": *I left explicit instructions. Implicit* means "implied, unstated": *We had an implicit understanding.*

farther, further *Farther* refers to additional distance (*How much farther is it to the beach?*), and *further* refers to additional time, amount, or other abstract matters (*I don't want to discuss this any further*).

feel Avoid this word in place of *think* or *believe: She thinks* [not *feels*] *that the law should be changed.*

fewer, less *Fewer* refers to individual countable items (a plural count noun°), *less* to general amounts (a noncount noun,° always singular): *Skim milk has fewer calories than whole milk. We have less milk left than I thought.*

further See *farther, further.*

get *Get* is easy to overuse; watch out for it in expressions such as *it's getting better* (substitute *improving*), *we got done* (substitute *finished*), and *the mayor has got to* (substitute *must*).

good, well *Good* is an adjective,° and *well* is nearly always an adverb:° *Larry's a good dancer. He and Linda dance well together. Well* is properly used as an adjective only to refer to health: *You look well.* (*You look good,* in contrast, means "Your appearance is pleasing.") See also p. 53.

hanged, hung Though both are past-tense forms° of *hang, hanged* is used to refer to executions and *hung* is used for all other meanings: *Tom Dooley was hanged* [not *hung*] *from a white oak tree. I hung* [not *hanged*] *the picture you gave me.*

have, of Use *have,* not *of,* after helping verbs° such as *could, should, would, may,* and *might: You should have* [not *should of*] *told me.*

he, she; he/she Convention has allowed the use of *he* to mean "he or she," but most writers today consider this usage inaccurate and unfair because it excludes females. The construction *he/she,* one substitute for *he,* is awkward and objectionable to many readers. The better choice is to recast the sentence in the plural, to rephrase, or to use *he or she*. For instance: *After infants learn to creep, they progress to crawling. After learning to creep, the infant progresses to crawling. After* *the infant learns to creep, he or she progresses to crawling.* See also pp. 27 and 50.

herself, himself See *myself, herself, himself, yourself.*

hisself Nonstandard for *himself.*

hopefully *Hopefully* means "with hope": *Freddy waited hopefully.* The use of *hopefully* to mean "it is to be hoped," "I hope," or "let's hope" is now very common; but since many readers continue to object strongly to the usage, you should avoid it. *I hope* [not *Hopefully*] *Eliza will be here soon.*

idea, ideal An *idea* is a thought or conception. An *ideal* (noun°) is a model of perfection or a goal. *Ideal* should not be used in place of *idea: The idea* [not *ideal*] *of the play is that our ideals often sustain us.*

if, whether For clarity, use *whether* rather than *if* when you are expressing an alternative: *If I laugh hard, people can't tell whether I'm crying.*

illicit See *elicit, illicit.*

illusion See *allusion, illusion.*

immigrate See *emigrate, immigrate.*

implicit See *explicit, implicit.*

imply, infer Writers or speakers *imply,* meaning "suggest": *Jim's letter implies he's having a good time.* Readers or listeners *infer,* meaning "conclude": *From Jim's letter I infer he's having a good time.*

irregardless Nonstandard for *regardless.*

is, are See *are, is.*

is when, is where These are faulty constructions in sentences that define: *Adolescence is a stage* [not *is when a person is*] *between childhood and adulthood. Socialism is a system in which* [not *is where*] *government owns the means of production.*

its, it's *Its* is the pronoun° *it* in the possessive case:° *That plant is losing its leaves. It's* is a contraction for *it is: It's likely to die if you don't water it.* See also pp. 78–79.

kind of, sort of, type of In formal speech and writing, avoid using *kind of* or *sort of* to mean "somewhat": *He was rather* [not *kind of*] *tall.*

Kind, sort, and *type* are singular: *This kind of dog is easily trained.* Errors often occur when these singular nouns are combined with the plural adjectives° *these* and *those: These kinds* [not *kind*] *of dogs are easily trained. Kind, sort,* and *type* should be followed by *of* but not by *a: I don't know what type of* [not *type* or *type of a*] *dog that is.*

Use *kind of, sort of,* or *type of* only when the word *kind, sort,* or *type* is important: *That was a strange* [not *strange sort of*] *statement.*

lay, lie *Lay* means "put" or "place" and takes a direct object:° *We could lay the tablecloth in the sun.* Its main forms are *lay, laid, laid. Lie* means "recline" or "be situated" and does not take an object: *I lie awake at night. The town lies east of the river.* Its main forms are *lie, lay, lain.*

less See *fewer, less.*

lie, lay See *lay, lie.*

like, as In formal speech and writing, *like* should not introduce a full clause.° The preferred choice is *as* or *as if: The plan succeeded as* [not *like*] *we hoped.* Use *like* only before a word or phrase: *Other plans like it have failed.*

literally This word means "actually" or "just as the words say," and it should not be used to intensify expressions whose words are not to be taken at face value. The sentence *He was literally climbing the walls* describes a person behaving like an insect, not a person who is restless or anxious. For the latter meaning, *literally* should be omitted.

lose, loose *Lose* means "mislay": *Did you lose a brown glove? Loose* usually means "unrestrained" or "not tight": *Ann's canary got loose.*

may, can See *can, may.*

may be, maybe *May be* is a verb,° and *maybe* is an adverb° meaning "perhaps": *Tuesday may be a legal holiday. Maybe we won't have classes.*

may of See *have, of.*

media *Media* is the plural of *medium* and takes a plural verb.° *All the news media are increasingly visual.* The singular verb is common, even in the media, but many readers prefer the plural verb and it is always correct.

might of See *have, of.*

must of See *have, of.*

myself, herself, himself, yourself The *-self* pronouns° refer to or intensify another word or words: *Paul did it himself; Jill herself said so.* In formal speech or writing, avoid using the *-self* pronouns in place of personal pronouns:° *No one except me* [not *myself*] *saw the accident. Michiko and I* [not *myself*] *planned the ceremony.*

never, no one See *all, always, never, no one.*

nowheres Nonstandard for *nowhere.*

number See *amount, number.*

of, have See *have, of.*

OK, O.K., okay All three spellings are acceptable, but avoid this colloquial term in formal speech and writing.

people, persons Except when emphasizing individuals, prefer *people* to *persons: We the people of the United States . . . ; Will the person or persons who saw the accident please notify. . . .*

percent (per cent), percentage Both these terms refer to fractions of one hundred. *Percent* always follows a numeral

(*40 percent of the voters*), and the word should be used instead of the symbol (%) in nontechnical writing. *Percentage* usually follows an adjective (*a high percentage*).

persons See *people, persons.*

phenomena The plural of *phenomenon* (meaning "perceivable fact" or "unusual occurrence"): *Many phenomena are not recorded. One phenomenon is attracting attention.*

plus *Plus* is standard as a preposition° meaning "in addition to": *His income plus mine is sufficient.* But *plus* is colloquial as a conjunctive adverb:° *Our organization is larger than theirs; moreover* [not *plus*], *we have more money.*

precede, proceed *Precede* means "come before": *My name precedes yours in the alphabet. Proceed* means "move on": *We were told to proceed to the waiting room.*

prejudice, prejudiced *Prejudice* is a noun;° *prejudiced* is an adjective.° Do not drop the *-d* from *prejudiced: I knew that my grandparents were prejudiced* [not *prejudice*].

principal, principle *Principal* is an adjective° meaning "foremost" or "major," a noun° meaning "chief official," or, in finance, a noun meaning "capital sum." *Principle* is a noun only, meaning "rule" or "axiom." *Her principal reasons for confessing were her principles of right and wrong.*

proceed, precede See *precede, proceed.*

raise, rise *Raise* means "lift" or "bring up" and takes a direct object:° *The Kirks raise cattle.* Its main forms are *raise, raised, raised. Rise* means "get up" and does not take an object: *They must rise at dawn.* Its main forms are *rise, rose, risen.*

real, really In formal speech and writing, *real* should not be used as an adverb;° *really* is the adverb and *real* an adjective.° *Popular reaction to the announcement was really* [not *real*] *enthusiastic.*

reason is because Although colloquially common, this construction should be avoided in formal speech and writing. Use a *that* clause after *reason is: The reason he is absent is that* [not *is because*] *he is sick.* Or: *He is absent because he is sick.*

respectful, respective *Respectful* means "full of (or showing) respect": *Be respectful of other people. Respective* means "separate": *The French and the Germans occupied their respective trenches.*

rise, raise See *raise, rise.*

sensual, sensuous *Sensual* suggests sexuality; *sensuous* means "pleasing to the senses." *Stirred by the sensuous scent of meadow grass and flowers, Cheryl and Paul found their thoughts turning sensual.*

set, sit *Set* means "put" or "place" and takes a direct object:° *He sets the pitcher down.* Its main forms are *set, set, set.*

Sit means "be seated" and does not take an object: *She sits on the sofa.* Its main forms are *sit, sat, sat.*

should of See *have, of.*

since *Since* mainly relates to time: *I've been waiting since noon.* But *since* can also mean "because": *Since you ask, I'll tell you.* Revise sentences in which the word could have either meaning, such as *Since you left, my life is empty.*

sit, set See *set, sit.*

somebody, some body; someone, some one *Somebody* and *someone* are indefinite pronouns;° *some body* is a noun° modified by *some;* and *some one* is a pronoun° or an adjective° modified by *some. Somebody ought to invent a shampoo that will give hair some body. Someone told James he should choose some one plan and stick with it.*

somewheres Nonstandard for *somewhere.*

sort of, sort of a See *kind of, sort of, type of.*

supposed to, used to In both these expressions, the *-d* is essential: *I used to* [not *use to*] *think so. He's supposed to* [not *suppose to*] *meet us.*

sure and, sure to; try and, try to *Sure to* and *try to* are the correct forms: *Be sure to* [not *sure and*] *buy milk. Try to* [not *Try and*] *find some decent tomatoes.*

take, bring See *bring, take.*

than, then *Than* is a conjunction° used in comparisons, *then* an adverb° indicating time: *Holmes knew then that Moriarty was wilier than he had thought.*

that, which *That* always introduces essential° clauses: *Use the lettuce that Susan bought* (the clause limits *lettuce* to a particular lettuce). *Which* can introduce both essential and nonessential° clauses, but many writers reserve *which* only for nonessential clauses: *The leftover lettuce, which is in the refrigerator, would make a good salad* (the clause simply provides more information about the lettuce we already know of). Essential clauses are not set off by commas; nonessential clauses are. See also pp. 70–71.

that, who, which Use *that* to refer to most animals and to things: *The animals that escaped included a zebra. The rocket that failed cost millions.* Use *who* to refer to people and to animals with names: *Dorothy is the girl who visits Oz. Her dog, Toto, who accompanies her, gives her courage.* Use *which* only to refer to animals and things: *The river, which runs a thousand miles, empties into the Indian Ocean.*

their, there, they're *Their* is the possessive° form of *they: Give them their money. There* indicates place (*I saw her standing there*) or functions as an expletive° (*There is a hole behind you*). *They're* is a contraction° for *they are: They're going fast.*

theirselves Nonstandard for *themselves.*

then, than See *than, then.*

these, this *These* is plural; *this* is singular. *This pear is ripe, but these pears are not.*

these kind, these sort, these type, those kind See *kind of, sort of, type of.*

thru A colloquial spelling of *through* that should be avoided in all academic and business writing.

to, too, two *To* is a preposition;° *too* is an adverb° meaning "also" or "excessively"; and *two* is a number. *I too have been to Europe two times.*

toward, towards Both are acceptable, though *toward* is preferred. Use one or the other consistently.

try and, try to See *sure and, sure to; try and, try to.*

type of See *kind of, sort of, type of.*

uninterested See *disinterested, uninterested.*

unique *Unique* means "the only one of its kind" and so cannot sensibly be modified with words such as *very* or *most: That was a unique* [not *a very unique* or *the most unique*] *movie.*

used to See *supposed to, used to.*

weather, whether The *weather* is the state of the atmosphere. *Whether* introduces alternatives. *The weather will determine whether we go or not.*

well See *good, well.*

whether, if See *if, whether.*

which, that See *that, which.*

who, which, that See *that, who, which.*

who, whom *Who* is the subject of a sentence or clause:° *We know who will come. Whom* is the object° of a verb° or preposition:° *We know whom we invited.*

who's, whose *Who's* is the contraction° of *who is: Who's at the door? Whose* is the possessive° form of *who: Whose book is that?*

would have Avoid this construction in place of *had* in clauses that begin *if: If the tree had* [not *would have*] *withstood the fire, it would have been the oldest in town.*

would of See *have, of.*

you In all but very formal writing, *you* is generally appropriate as long as it means "you, the reader." In all writing, avoid indefinite uses of *you,* such as *In one ancient tribe your first loyalty was to your parents.*

your, you're *Your* is the possessive° form of *you: Your dinner is ready. You're* is the contraction° of *you are: You're bound to be late.*

yourself See *myself, herself, himself, yourself.*

Glossary of Terms

This glossary defines the terms and concepts of basic English grammar, including every term marked ° in the text.

absolute phrase A phrase that consists of a noun° or pronoun° plus the *-ing* or *-ed* form of a verb° (a participle°): *Our accommodations arranged, we set out on our trip. They will hire a local person, other things being equal.*

active voice The verb form° used when the sentence subject° names the performer of the verb's action: *The drillers used a rotary blade.* For more, see *voice.*

adjective A word used to modify a noun° or pronoun:° *beautiful morning, ordinary one, good spelling.* Contrast *adverb.* Nouns, word groups, and some verb° forms may also serve as adjectives: *book sale; sale of old books; the sale, which occurs annually; increasing profits.*

adverb A word used to modify a verb,° an adjective,° another adverb, or a whole sentence: *warmly greet* (verb), *only three people* (adjective), *quite seriously* (adverb), *Fortunately, she is employed* (sentence). Word groups may also serve as adverbs: *drove by a farm, plowed the fields when the earth thawed.*

agreement The correspondence of one word to another in person,° number,° or gender.° Mainly, a verb° must agree with its subject° (*The chef orders eggs*), and a pronoun° must agree with its antecedent° (*The chef surveys her breakfast*). See also pp. 43–46 and 49–51.

antecedent The word a pronoun° refers to: *Jonah, who is not yet ten, has already chosen the college he will attend* (*Jonah* is the antecedent of the pronouns *who* and *he*).

appositive A word or word group appearing next to a noun° or pronoun° that renames or identifies it and is equivalent to it: *My brother Michael, the best horn player in town, won the state competition* (*Michael* identifies which brother is being referred to; *the best horn player in town* renames *My brother Michael*).

article The words *a, an,* and *the.* A kind of determiner,° an article always signals that a noun follows. See p. 219 for how to choose between *a* and *an.* See pp. 56–58 for the rules governing *a/an* and *the.*

auxiliary verb See *helping verb.*

case The form of a pronoun° or noun° that indicates its function in the sentence. Most pronouns have three cases. The **subjective case** is for subjects° and subject complements:° *I, you, he, she, it, we, they, who, whoever.* The **objective case** is for objects:° *me, you, him, her, it, us, them, whom, whomever.* The **possessive case** is for ownership: *my/mine, your/yours, his, her/hers, its, our/ours, their/theirs, whose.* Nouns use the subjective form (*dog, America*) for all cases except the possessive (*dog's, America's*).

clause A group of words containing a subject° and a predicate.° A **main clause** can stand alone as a sentence: *We can go to the movies.* A **subordinate clause** cannot stand alone as a sentence: *We can go if Julie gets back on time.* For more, see *subordinate clause.*

collective noun A word with singular form that names a group of individuals or things: for instance, *team, army, family, flock, group.* A collective noun generally takes a singular verb and a singular pronoun: *The army is prepared for its role.* See also pp. 45 and 51.

comma splice A sentence error in which two sentences (main clauses°) are separated by a comma without *and, but, or, nor,* or another coordinating conjunction.° Splice: *The book was long, it contained useful information.* Revised: *The book was long; it contained useful information.* Or: *The book was long, and it contained useful information.* See pp. 63–65.

comparison The form of an adverb° or adjective° that shows its degree of quality or amount. The **positive** is the simple, uncompared form: *gross, clumsily.* The **comparative** compares the thing modified to at least one other thing: *grosser, more clumsily.* The **superlative** indicates that the thing modified exceeds all other things to which it is being compared: *grossest, most clumsily.* The comparative and superlative are formed either with the endings *-er* and *-est* or with the words *more* and *most* or *less* and *least.*

complement See *subject complement.*

complex sentence See *sentence.*

compound-complex sentence See *sentence.*

compound construction Two or more words or word groups serving the same function, such as a compound subject° (*Harriet and Peter poled their barge down the river*), a compound predicate° (*The scout watched and waited*), or a compound sentence° (*He smiled, and I laughed*).

compound sentence See *sentence.*

conditional statement A statement expressing a condition contrary to fact and using the subjunctive mood° of the verb: *If she were mayor, the unions would cooperate.*

conjunction A word that links and relates parts of a sentence. See *coordinating conjunction* (*and, but,* etc.), *correlative conjunction* (*either . . . or, both . . . and,* etc.), and *subordinating conjunction* (*because, if,* etc.).

conjunctive adverb An adverb° that can relate two main clauses° in a single sentence: *We had hoped to own a house by now; however, prices are still too high.* The main clauses are separated by a semicolon or a period. Some common conjunctive adverbs: *accordingly, also, anyway, besides, certainly, consequently, finally, further, furthermore, hence, however, incidentally, indeed, instead, likewise, meanwhile, moreover, namely, nevertheless, next, nonetheless, now, otherwise, similarly, still, then, thereafter, therefore, thus, undoubtedly.*

contraction A condensed expression, with an apostrophe replacing the missing letters: for example, *doesn't* (*does not*), *we'll* (*we will*).

coordinating conjunction A word linking words or word groups serving the same function: *The dog and cat sometimes fight, but they usually get along.* The coordinating conjunctions are *and, but, or, nor, for, so, yet.*

coordination The linking of words or word groups that are of equal importance, usually with a coordinating conjunction.° *He and I laughed, but she was not amused.* Contrast *subordination.*

correlative conjunction Two or more connecting words that work together to link words or word groups serving the same function: *Both Michiko and June signed up, but neither Stan nor Carlos did.* The correlatives include *both . . . and, just as . . . so, not only . . . but also, not . . . but, either . . . or, neither . . . nor, whether . . . or, as . . . as.*

count noun A word that names a person, place, or thing that can be counted (and so may appear in plural form): *camera/cameras, river/rivers, child/children.*

dangling modifier A modifier that does not sensibly describe anything in its sentence. Dangling: *Having arrived late, the concert had already begun.* Revised: *Having arrived late, we found that the concert had already begun.* See p. 60.

determiner A word such as *a, an, the, my,* and *your* that indicates that a noun follows. See also *article.*

direct address A construction in which a word or phrase indicates the person or group spoken to: *Have you finished, John? Farmers, unite.*

direct object A noun° or pronoun° that identifies who or what receives the action of a verb:° *Education opens doors.* For more, see *object* and *predicate.*

direct question A sentence asking a question and concluding with a question mark: *Do they know we are watching?* Contrast *indirect question.*

direct quotation Repetition of what someone has written or said, using the exact words of the original and enclosing them in quotation marks: *Feinberg writes, "The reasons are both obvious and sorry."*

double negative A nonstandard form consisting of two negative words used in the same construction so that they effectively cancel each other: *I don't have no money.* Rephrase as *I have no money* or *I don't have any money.* See also p. 55.

ellipsis The omission of a word or words from a quotation, indicated by the three spaced periods of an **ellipsis mark:** *"all . . . are created equal."* See also pp. 85–87.

essential element A word or word group that is essential to the meaning of the sentence because it limits the word it refers to: removing it would leave the meaning unclear or too general. Essential elements are *not* set off by commas: *Dorothy's companion the Scarecrow lacks a brain. The man who called about the apartment said he'd try again.* Contrast *nonessential element.* See also pp. 70–71.

expletive construction A sentence that postpones the subject° by beginning with *there* or *it* and a form of *be: It is impossible to get a ticket. There are no more seats available.*

first person See *person.*

fused sentence (run-on sentence) A sentence error in which two complete sentences (main clauses°) are joined with no punctuation or connecting word between them. Fused: *I heard his lecture it was dull.* Revised: *I heard his lecture; it was dull.* See pp. 63–65.

future perfect tense The verb tense expressing an action that will be completed before another future action: *They will have heard by then.* For more, see *tense.*

future tense The verb tense expressing action that will occur in the future: *They will hear soon.* For more, see *tense.*

gender The classification of nouns° or pronouns° as masculine (*he, boy*), feminine (*she, woman*), or neuter (*it, computer*).

generic *he* *He* used to mean *he or she.* Avoid *he* when you intend either or both genders. See pp. 27 and 50.

generic noun A noun° that does not refer to a specific person or thing: *Any person may come. A student needs good work habits. A school with financial problems may shortchange its students.* A singular generic noun takes a singular pronoun° (*he, she,* or *it*). See also *indefinite pronoun* and p. 50.

gerund A verb form that ends in *-ing* and functions as a noun:° *Working is all right for killing time.* For more, see *verbals and verbal phrases.*

gerund phrase See *verbals and verbal phrases.*

helping verb (auxiliary verb) A verb° used with another verb to convey time, possibility, obligation, and other meanings: *You should write a letter. You have written other letters.* The **modals** are the following: *be able to, be supposed to, can, could, had better, had to, may, might, must, ought to, shall, should, used to, will, would.* The other helping verbs are forms of *be, have,* and *do.* See also pp. 34–36.

idiom An expression that is peculiar to a language and that may not make sense if taken literally: for example, *bide your time, by and large,* and *put up with.*

imperative See *mood.*

indefinite pronoun A word that stands for a noun° and does not refer to a specific person or thing. A few indefinite pronouns are plural (*both, few, many, several*) or may be singular or plural (*all, any, more, most, some*). But most are only singular: *anybody, anyone, anything, each, either, everybody, everyone, everything, neither, nobody, none, no one, nothing, one, somebody, someone, something.* The singular indefinite pronouns take singular verbs and are referred to by singular pronouns: *Something makes its presence felt.* See also *generic noun* and pp. 44 and 50.

indicative See *mood.*

indirect object A noun° or pronoun° that identifies to whom or what something is done: *Give them the award.* For more, see *object* and *predicate.*

indirect question A sentence reporting a question and ending with a period: *Writers wonder if their work must always be lonely.* Contrast *direct question.*

indirect quotation A report of what someone has written or said, but not using the exact words of the original and not enclosing the words in quotation marks. Quotation: *"Events have controlled me."* Indirect quotation: *Lincoln said that events had controlled him.*

infinitive A verb form° consisting of the verb's dictionary form plus *to: to swim, to write.* For more, see *verbals and verbal phrases.*

infinitive phrase See *verbals and verbal phrases.*

intensive pronoun See *pronoun.*

interjection A word standing by itself or inserted in a construction to exclaim: *Hey! What the heck did you do that for?*

interrogative pronoun A word that begins a question and serves as the subject° or object° of the sentence. The interrogative pronouns are *who, whom, whose, which,* and *what. Who received the flowers? Whom are they for?*

intransitive verb A verb° that does not require a following word (direct object°) to complete its meaning: *Mosquitoes buzz. The hospital may close.* For more, see *predicate.*

irregular verb See *verb forms.*

linking verb A verb that links, or connects, a subject° and a word that renames or describes the subject (a subject complement°): *They are golfers. You seem lucky.* The linking verbs are the forms of *be,* the verbs of the senses (*look, sound, smell, feel, taste*), and a few others (*appear, become, grow, prove, remain, seem, turn*). For more, see *predicate.*

main clause A word group that contains a subject° and a predicate,° does not begin with a subordinating word, and may stand alone as a sentence: *The president was not overbearing.* For more, see *clause.*

main verb The part of a verb phrase° that carries the principal meaning: *had been walking, could happen, was chilled.* Contrast *helping verb.*

misplaced modifier A modifier whose position makes unclear its relation to the rest of the sentence. Misplaced: *The children played with firecrackers that they bought illegally in the field.* Revised: *The children played in the field with firecrackers that they bought illegally.*

modal See *helping verb.*

modifier Any word or word group that limits or qualifies the meaning of another word or word group. Modifiers include adjectives° and adverbs° as well as words and word groups that act as adjectives and adverbs.

mood The form of a verb° that shows how the speaker views the action. The **indicative mood**, the most common, is used to make statements or ask questions: *The play will be performed Saturday. Did you get tickets?* The **imperative mood** gives a command: *Please get good seats. Avoid the top balcony.* The **subjunctive mood** expresses a wish, a condition contrary to fact, a recommendation, or a request: *I wish George were coming with us. If he were here, he'd come. I suggested that he come. The host asked that he be here.*

noncount noun A word that names a person, place, or thing and that is not considered countable in English (and so does not appear in plural form): *confidence, information, silver, work.* See pp. 57–58 for a longer list.

nonessential element A word or word group that does not limit the word it refers to and that is not essential to the meaning of the sentence. Nonessential elements are usually set off by commas: *Sleep, which we all need, occupies a third of our lives. His wife, Patricia, is a chemist.* Contrast *essential element.* See also pp. 70–71.

nonrestrictive element See *nonessential element.*

noun A word that names a person, place, thing, quality, or idea: *Maggie, Alabama, clarinet, satisfaction, socialism.* See also *collective noun, count noun, generic noun, noncount noun,* and *proper noun.*

noun clause See *subordinate clause.*

number The form of a word that indicates whether it is singular or plural. Singular: *I, he, this, child, runs, hides.* Plural: *we, they, these, children, run, hide.*

object A noun° or pronoun° that receives the action of or is influenced by another word. A **direct object** receives the action of a verb° or verbal° and usually follows it: *We watched* *the stars.* An **indirect object** tells for or to whom something is done: *Reiner bought us tapes.* An **object of a preposition** usually follows a preposition:° *They went to New Orleans.*

objective case The form of a pronoun° when it is the object° of a verb° (*call him*) or the object of a preposition° (*for us*). For more, see *case.*

object of preposition See *object.*

parallelism Similarity of form between two or more coordinated elements: *Rising prices and declining incomes left many people in bad debt and worse despair.* See also pp. 17–19.

parenthetical expression A word or construction that interrupts a sentence and is not part of its main structure, called *parenthetical* because it could (or does) appear in parentheses: *Mary Cassatt (1845–1926) was an American painter. Her work, incidentally, is in the museum.*

participial phrase See *verbals and verbal phrases.*

participle See *verbals and verbal phrases.*

particle A preposition° or adverb° in a two-word verb: *catch on, look up.*

parts of speech The classes of words based on their form, function, and meaning: nouns, pronouns, verbs, adjectives, adverbs, conjunctions, prepositions, and interjections. See separate entries for each part of speech.

passive voice The verb form° used when the sentence subject° names the receiver of the verb's action: *The mixture was stirred.* For more, see *voice.*

past participle The *-ed* form of most verbs:° *fished, hopped.* The past participle may be irregular: *begun, written.* For more, see *verbals and verbal phrases* and *verb forms.*

past perfect tense The verb tense expressing an action that was completed before another past action: *No one had heard that before.* For more, see *tense.*

past tense The verb tense expressing action that occurred in the past: *Everyone laughed.* For more, see *tense.*

past-tense form The verb form used to indicate action that occurred in the past, usually created by adding *-d* or *-ed* to the verb's dictionary form (*smiled*) but created differently for most irregular verbs (*began, threw*). For more, see *verb forms.*

perfect tenses The verb tenses indicating action completed before another specific time or action: *have walked, had walked, will have walked.* For more, see *tense.*

person The form of a verb° or pronoun° that indicates whether the subject is speaking, spoken to, or spoken about. In the **first person** the subject is speaking: *I am, we are.* In the **second person** the subject is spoken to: *you are.* In the **third person** the subject is spoken about: *he/she/it is, they are.*

personal pronoun *I, you, he, she, it, we,* or *they:* a word that substitutes for a specific noun° or other pronoun. For more, see *case.*

phrase A group of related words that lacks a subject° or a predicate° or both: *She ran into the field. She tried to jump the fence.* See also *absolute phrase, prepositional phrase, verbals and verbal phrases.*

plain form The dictionary form of a verb: *buy, make, run, swivel.* For more, see *verb forms.*

plural More than one. See *number.*

positive form See *comparison.*

possessive case The form of a noun° or pronoun° that indicates its ownership of something else: *men's attire, your briefcase.* For more, see *case.*

predicate The part of a sentence that makes an assertion about the subject.° The predicate may consist of an intransitive verb° (*The earth trembled*), a transitive verb° plus direct object° (*The earthquake shook buildings*), a linking verb° plus subject complement° (*The result was chaos*), a transitive verb plus indirect object° and direct object (*The government sent the city aid*), or a transitive verb plus direct object and object complement (*The citizens considered the earthquake a disaster*).

preposition A word that forms a noun° or pronoun° (plus any modifiers) into a **prepositional phrase**: *about love, down the steep stairs.* The common prepositions: *about, above, according to, across, after, against, along, along with, among, around, as, at, because of, before, behind, below, beneath, beside, between, beyond, by, concerning, despite, down, during, except, except for, excepting, for, from, in, in addition to, inside, in spite of, instead of, into, like, near, next to, of, off, on, onto, out, out of, outside, over, past, regarding, since, through, throughout, till, to, toward, under, underneath, unlike, until, up, upon, with, within, without.*

prepositional phrase A word group consisting of a preposition° and its object.° Prepositional phrases usually serve as adjectives° (*We saw a movie about sorrow*) and as adverbs° (*We went back for the second show*).

present participle The *-ing* form of a verb:° *swimming, flying.* For more, see *verbals and verbal phrases.*

present perfect tense The verb tense expressing action that began in the past and is linked to the present: *Dogs have buried bones here before.* For more, see *tense.*

present tense The verb tense expressing action that is occurring now, occurs habitually, or is generally true: *Dogs bury bones here often.* For more, see *tense.*

principal parts The three forms of a verb from which its various tenses are created: the **plain form** (*stop, go*), the **past-**

tense form (*stopped, went*), and the **past participle** (*stopped, gone*). For more, see *tense* and *verb forms.*

progressive tenses The verb tenses that indicate continuing (progressive) action and use the *-ing* form of the verb: *A dog was barking here this morning.* For more, see *tense.*

pronoun A word used in place of a noun,° such as *I, he, everyone, who,* and *herself.* See also *indefinite pronoun, interrogative pronoun, personal pronoun, relative pronoun.*

proper adjective A word formed from a proper noun° and used to modify a noun° or pronoun:° *Alaskan winter.*

proper noun A word naming a specific person, place, or thing and beginning with a capital letter: *David Letterman, Mt. Rainier, Alaska, US Congress.*

regular verb See *verb forms.*

relative pronoun A word that relates a group of words to a noun° or another pronoun.° The relative pronouns are *who, whom, whoever, whomever, which,* and *that. Ask the woman who knows all. This may be the question that stumps her.* For more, see *case.*

restrictive element See *essential element.*

run-on sentence See *fused sentence.*

-s form See *verb forms.*

second person See *person.*

sentence A complete unit of thought, consisting of at least a subject° and a predicate° that are not introduced by a subordinating word. A **simple sentence** contains one main clause:° *I'm leaving.* A **compound sentence** contains at least two main clauses: *I'd like to stay, but I'm leaving.* A **complex sentence** contains one main clause and at least one subordinate clause:° *If you let me go now, you'll be sorry.* A **compound-complex sentence** contains at least two main clauses and at least one subordinate clause: *I'm leaving because you want me to, but I'd rather stay.*

sentence fragment An error in which an incomplete sentence is set off as a complete sentence. Fragment: *She was not in shape for the race. Which she had hoped to win.* Revised: *She was not in shape for the race, which she had hoped to win.* See pp. 61–63.

series Three or more items with the same function: *We gorged on ham, eggs, and potatoes.*

simple sentence See *sentence.*

simple tenses See *tense.*

singular One. See *number.*

split infinitive The usually awkward interruption of an infinitive° and its marker *to* by a modifier: *Management decided to not introduce the new product.* See p. 59.

squinting modifier A modifier that could modify the words on either side of it: *The plan we considered seriously worries me.*

subject In grammar, the part of a sentence that names something and about which an assertion is made in the predicate:° *The quick, brown fox jumped lazily* (simple subject); *The quick, brown fox jumped lazily* (complete subject).

subject complement A word that renames or describes the subject° of a sentence, after a linking verb.° *The stranger was a man* (noun°). *He seemed gigantic* (adjective°).

subjective case The form of a pronoun° when it is the subject° of a sentence (*I called*) or a subject complement° (*It was I*). For more, see *case.*

subjunctive See *mood.*

subordinate clause A word group that consists of a subject° and a predicate,° begins with a subordinating word such as *because* or *who,* and is not a question: *They voted for whoever cared the least because they mistrusted politicians.* Subordinate clauses may serve as adjectives° (*The car that hit Fred was blue*), as adverbs° (*The car hit Fred when it ran a red light*), or as nouns° (*Whoever was driving should be arrested*). Subordinate clauses are *not* complete sentences.

subordinating conjunction A word that turns a complete sentence into a word group (a subordinate clause°) that can serve as an adverb° or a noun.° *Everyone was relieved when the meeting ended.* Some common subordinating conjunctions: *after, although, as, as if, as long as, as though, because, before, even if, even though, if, if only, in order that, now that, once, rather than, since, so that, than, that, though, till, unless, until, when, whenever, where, whereas, wherever, while.*

subordination Deemphasizing one element in a sentence by making it dependent on rather than equal to another element. Through subordination, *I left six messages; the doctor failed to call* becomes *Although I left six messages, the doctor failed to call* or *After six messages, the doctor failed to call.*

superlative See *comparison.*

tag question A question attached to the end of a statement and composed of a pronoun,° a helping verb,° and sometimes the word *not: It isn't raining, is it? It is sunny, isn't it?*

tense The verb form that expresses time, usually indicated by endings and by helping verbs. See also *verb forms.*

Present Action that is occurring now, occurs habitually, or is generally true

Simple present Plain form or -*s* form	Present progressive *Am, is,* or *are* plus -*ing* form
I *walk.*	I *am walking.*
You/we/they *walk.*	You/we/they *are walking.*
He/she/it *walks.*	He/she/it *is walking.*

Past Action that occurred before now

Simple past Past-tense form	Past progressive *Was* or *were* plus *-ing* form
I/he/she/it *walked.* You/we/they *walked.*	I/he/she/it *was walking.* You/we/they *were walking.*

Future Action that will occur in the future

Simple future *Will* plus plain form	Future progressive *Will be* plus *-ing* form
I/you/he/she/it/we/they *will walk.*	I/you/he/she/it/we/they *will be walking.*

Present perfect Action that began in the past and is linked to the present

Present perfect *Have* or *has* plus past participle	Present perfect progressive *Have been* or *has been* plus *-ing* form
I/you/we/they *have walked.*	I/you/we/they *have been walking.*
He/she/it *has walked.*	He/she/it *has been walking.*

Past perfect Action that was completed before another past action

Past perfect *Had* plus past participle	Past perfect progressive *Had been* plus *-ing* form
I/you/he/she/it/we/they *had walked.*	I/you/he/she/it/we/they *had been walking.*

Future perfect Action that will be completed before another future action

Future perfect *Will have* plus past participle	Future perfect progressive *Will have been* plus *-ing* form
I/you/he/she/it/we/they *will have walked.*	I/you/he/she/it/we/they *will have been walking.*

third person See *person*.

transitional expression A word or phrase that links sentences and shows the relations between them. Transitional expressions can signal various relationships (examples in parentheses): addition or sequence (*also, besides, finally, first, furthermore, in addition, last*); comparison (*also, similarly*); contrast (*even so, however, in contrast, still*); examples (*for example, for instance, that is*); intensification (*indeed, in fact, of course*); place (*below, elsewhere, here, nearby, to the east*); time (*afterward, at last, earlier, immediately, meanwhile, simultaneously*); repetition or summary (*in brief, in other words, in short, in summary, that is*); and cause and effect (*as a result, consequently, hence, therefore, thus*).

transitive verb A verb° that requires a following word (a direct object°) to complete its meaning: *We raised the roof.* For more, see *predicate*.

verb A word that expresses an action (*bring, change*), an occurrence (*happen, become*), or a state of being (*be, seem*). A verb is the essential word in a predicate,° the part of a sentence that makes an assertion about the subject.° With endings and helping verbs,° verbs can indicate tense,° mood,° voice,° number,° and person.° For more, see separate entries for each of these aspects as well as *verb forms*.

verbals and verbal phrases **Verbals** are verb forms used as adjectives,° adverbs,° or nouns.° They form **verbal phrases** with objects° and modifiers.° A **present participle** adds *-ing* to the dictionary form of a verb (*living*). A **past participle** usually adds *-d* or *-ed* to the dictionary form (*lived*), although irregular verbs work differently (*begun, swept*). A participle or **participial phrase** usually serves as an adjective: *Strolling shoppers fill the malls.* A **gerund** is the *-ing* form of a verb used as a noun. Gerunds and **gerund phrases** can do whatever nouns can do: *Shopping satisfies needs.* An **infinitive** is the verb's dictionary form plus *to*: *to live.* Infinitives and **infinitive phrases** may serve as nouns (*To design a mall is a challenge*), as adverbs (*Malls are designed to make shoppers feel safe*), or as adjectives (*The mall supports the impulse to shop*).

A verbal *cannot* serve as the only verb in a sentence. For that, it requires a helping verb:° *Shoppers were strolling.*

verb forms Verbs have five distinctive forms. The **plain form** is the dictionary form: *A few artists live in town today.* The **-s form** adds *-s* or *-es* to the plain form: *The artist lives in town today.* The **past-tense form** usually adds *-d* or *-ed* to the plain form: *Many artists lived in town before this year.* Some verbs' past-tense forms are irregular, such as *began, fell, swam, threw, wrote.* The **past participle** is usually the same as the past-tense form, although, again, some verbs' past participles are irregular (*begun, fallen, swum, thrown, written*). The **present participle** adds *-ing* to the plain form: *A few artists are living in town today.*

Regular verbs are those that add *-d* or *-ed* to the plain form for the past-tense form and past participle. **Irregular verbs** create these forms in irregular ways (see above).

verb phrase A verb° of more than one word that serves as the predicate° of a sentence: *The movie has started.*

voice The form of a verb° that tells whether the sentence subject° performs the action or is acted upon. In the **active voice** the subject acts: *The city controls rents.* In the **passive voice** the subject is acted upon: *Rents are controlled by the city.* See also pp. 41–42.

Index

D

E

J

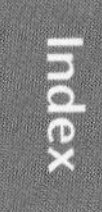
Index

O

Index

Detailed Contents